A KEY TO THE
ADULTS AND NYMPHS
OF THE BRITISH

Stoneflies
(Plecoptera)

with notes on their
Ecology and Distribution

by

H. B. N. HYNES, D.Sc.

Department of Biology
University of Waterloo, Ontario

FRESHWATER BIOLOGICAL ASSOCIATION
SCIENTIFIC PUBLICATION No. 17
Third edition, with minor revision, 1977
Reprinted 1984, 1993

FOREWORD

Dr Hynes has now given us a revised and enlarged key to replace the second of this series published in 1940. Since then the stoneflies have been much studied by Dr Hynes himself, by others in this country, and by Scandinavian and Continental workers, so that it is probable that more important papers on the group have been published in the last twenty years than in any previous comparable period. The keys to the adults have therefore required a good deal of revision and have also been somewhat rearranged in the light of experience. It has also been possible to include keys to the nymphs, all but one of which are now known. An innovation in the series is the addition of a series of maps illustrating the known distribution of each species in the British Isles.

THE FERRY HOUSE   H. C. GILSON
May 1957   *Director*

FOREWORD to the SECOND EDITION

A second edition has provided Dr Hynes with the opportunity to make one nomenclatorial revision and to insert several improvements to the key suggested from the experience of users; in particular the key to *Nemoura* has been significantly revised with the inclusion of new illustrations. The distribution maps have been brought up to date.

THE FERRY HOUSE   H. C. GILSON
August 1967   *Director*

SBN 900386 28 2

FOREWORD to the THIRD EDITION

This is a reprint of the second (1967) edition, with the addition of several references, and notes on the present status of *Chloroperla* and recent findings on the nymphs of *Isoperla*.

The distribution maps of the second edition are reproduced unchanged. It has not been found possible to update them or re-plot them on the 10 km grid, but it was felt that a number of interesting points regarding distribution would be lost if they were omitted.

THE FERRY HOUSE
February 1977

E. D. LE CREN
Director

CONTENTS

INTRODUCTION

The order Plecoptera was originally placed by Linnaeus as the family Perlidae in his order Neuroptera; it is now considered to be a separate order, as are most of the families that were placed there with it, for example the Odonata (dragonflies) and the Ephemeroptera (mayflies, etc.). The stoneflies form one of the most primitive groups of winged insects existing today. They are closely related to the fossil Paraplecoptera from the Carboniferous, and would seem to have been derived from the family Narkemidae of that order in the Lower Permian (Sharov 1960). They differ from their fossil ancestors in reduction of wing venation and in having only three tarsal segments. In the Upper Permian a representative of the modern Australian family Eustheniidae has been found; a similar instance of a Palaeozoic insect belonging to a modern family is known only among the Orthoptera (cockroaches, grasshoppers, etc.), which are generally considered to be the most primitive of all modern winged insects. The stoneflies are less primitive in some respects than the Ephemeroptera and Orthoptera. For instance, it is impossible to homologise their genitalia with structures in other orders, a fact which suggests that their genitalia are secondary developments; the wing venation is to some extent specialised, and there are only three tarsal joints instead of the primitive five.

There are three sub-orders, the primitive Archiperlaria, which includes the family Eustheniidae and is confined to the southern hemisphere. These gave rise to the Filipalpia, represented in Britain by the families Taeniopterygidae, Leuctridae, Capniidae and Nemouridae, and to the Setipalpia, represented in Britain by the Perlodidae, Perlidae and Chloroperlidae. During the course of their evolution stoneflies have apparently several times crossed the equator thus producing groups more or less isolated in the northern or southern hemispheres, and it is probably true to say that in no other order of insects is the evolutionary history so well understood (Illies 1965).

The keys present characters by which adults of both sexes, and nymphs longer than about 5 mm, may be identified to species. In compiling them every effort has been made to use characters which are easily seen, and, where possible, several are given. As many of these characters, e.g. adult genitalia, colour pattern, and arrangements of bristles on nymphs, are difficult to describe, the figures should always be used in conjunction with the keys. All the figures have been drawn from British specimens except where noted in the legends.

In adults of the small species the genitalia are often the only parts in which constant specific differences can be found, and these have been illustrated for all species. In certain species males with both short and long wings occur, but in others wing-length is constant. Aubert (1945, 1949) has discussed the problem of specific characters in the order.

Among nymphs of the smaller species odd specimens not infrequently occur which are atypical in one or more of the aspects which are otherwise characteristic of the species. This unfortunately can lead to misidentification no matter how carefully keys are prepared, and it is always better to base identification of nymphs on several specimens from the same locality than to study only one specimen, which may be aberrant.

STRUCTURE OF THE ADULT   The most important structures in the taxonomy of the stoneflies are the wings and the genitalia. Other features (see Fig. IA) are: the antennae (*a*); the proportions of the three segments of the tarsus (*t*); an M-shaped mark, the M-line (*ml*), which crosses the head in front of the eyes in some genera; the pronotum (*pn*) or dorsal plate of the first thoracic segment (prothorax); the vestiges of the nymphal tracheal gills (*g*) which occur in some genera (these may not be much reduced from the nymphal condition in newly emerged specimens); and the cerci (*c*), which are often reduced to one or very few segments. In the description of the genitalia the terms *tergum* and *sternum* (plural *terga* and *sterna*) refer to the dorsal and ventral plates respectively of the abdominal segment in question.

There has been much disagreement about the nomenclature of the various wing veins, particularly in the hind wing. The system followed here is that of Comstock (1918) and is based on a study of the developing nymphal wings. It differs very little from the more modern conception of Tillyard, the chief difference being that the latter regards the vein Cu I as primarily three branched. Comstock's system is used here because it is the system which has been used, with modifications, in most of the major European works on the group. The veins are as follows (see Fig. IA):

The costa (C) forms the front margin of the wing. The subcosta

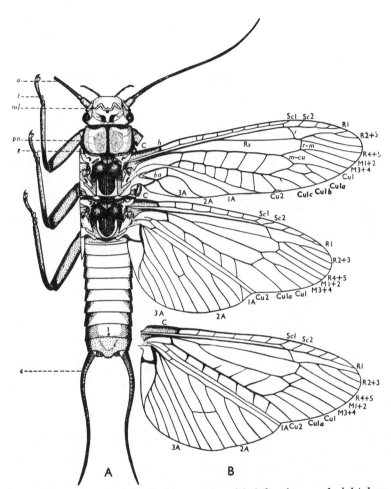

Fig. 1. A, female of *Perla bipunctata* with left wings and right legs removed. *a*, antenna; *t*, tarsus; *ml*, M-line; *pn*, pronotum; *g*, vestige of nymphal tracheal gill, × 3·5.
B, right hind-wing of female *Dinocras cephalotes* × 3·5.
For lettering of wing venation see text (p. 6).

(Sc) is two branched in the stoneflies, the branch Sc2 fusing for part of its length with the vein R1 behind it. The radius (R) has a front branch R1 and a hind branch, the radial sector (Rs); this has primitively four branches, but in the stoneflies they are reduced to two (R2+3 and R4+5), which may be secondarily re-divided. In the hind wing the radial sector is fused to the media in the adult, but is separate in the nymph, as has been found by Comstock. This fact has caused some European authors (Klapálek, Schoenemund, and Kühtreiber) to say that Rs is absent from the hind wing. The media (M) is two branched in the stoneflies (M1 + 2 and M3 + 4). The cubitus (Cu) is two branched, the branch Cu 1 bending forward towards and often fusing for part of its length with M3 + 4. Cu 1 often bears accessory veins, which are named Cu 1a, Cu 1b, etc. Of the anal veins (1A, 2A, and 3A) the hinder two are often branched; 1A is only rarely so.

In addition to the veins there are certain cross-veins which occur throughout the order, and are of importance. These are the humeral cross-vein (*h*), the radial cross-vein (*r*), the radio-medial cross-vein (*r-m*), the medio-cubital cross-vein (*m-cu*), and the anal cross-vein between 1A and 2A which cuts off the basal anal cell (*ba*). In fact the medio-cubital cross-vein is not a true cross-vein, but represents the posterior median vein: the median vein in Plecoptera, in contrast to most other orders, is the anterior median.

In the short-winged males the venation may be rather confused, but once the female venation is understood the male can usually be made out also (Fig. 6).

The genitalia are very diverse within the order, and are unlike those of any other group of insects. In the female the sternum of the 7th, 8th, or 9th abdominal segment is modified to form what is known as the sub-genital plate. When on the 7th segment, the sternum usually has a backwardly directed projection of the hind margin which lies over the 8th sternum; e.g. Nemouridae (Figs. 11 and 12), Capniidae (Fig. 14), Perlodidae (Figs. 16 and 17), Perlidae (Fig. 18) and Chloroperlidae (Fig. 19). Among the Nemouridae *Protonemura* (Fig. 10) is an exception in that it has no true sub-genital plate. In some species with the sub-genital plate on the 7th segment the 8th sternum is modified by thickening etc. When the sub-genital plate is on the 8th segment the sternum is modified in shape and may also be thickened in parts: e.g. Leuctridae (Fig. 13). When the sub-genital plate is on the 9th segment the sternum is often elongated and produced backwards over the 10th segment: e.g. Taeniopterygidae (Figs. 8 and 9). It will be seen therefore, that the term sub-genital plate covers three types of non-homologous structure. The female cerci may be long and many segmented (Perlodidae, Perlidae, Chloroperlidae and Capniidae), short and few segmented (Taeniopterygidae), or reduced to one segment, which may or may not bear vestiges of further segments at the tip (Leuctridae and Nemouridae).

The male genitalia are very varied. There are eight types in this country, which are described briefly below. In all, as in the females, two sclerites, the paraprocts, lie behind the 10th sternum, which, in several families, is very narrow and membranous. There is also a dorsal structure, the epiproct, behind the 10th tergum, which in many families is complex and projects upwards or forwards over the tergum.

1. Taeniopterygidae (Figs. 8 and 9). Cerci reduced to two or very few segments and directed dorsally; 10th tergum much modified, and the more anterior ones narrowed; 9th sternum extending backwards where it is upturned, forming a ventral plate (or sub-genital plate, but this term is best reserved for the female structure). On its anterior margin there is in most genera a small lobe, which is used in drumming on the substratum during courtship. Paraprocts small, complex, and asymmetrical.

2. Nemouridae (Figs. 10, 11 and 12). Cerci one-segmented, often modified, and in some species bearing remnants of further segments at the tip; 9th sternum drawn out posteriorly and bearing a drumming lobe anteriorly; 10th tergum partly covered by the anteriorly directed and complex epiproct. Paraprocts modified: in *Protonemura* and *Nemurella,* they are divided into three parts, a median lamella, a sub-anal plate and an external appendage which lies next to the cercus; in *Amphinemura* these three parts are also present, but remain fused to one another at the base.

3. *Leuctra* (Fig. 13). Cerci one-segmented, but often with remnants of further segments at the tip; 9th sternum bearing a small drumming lobe anteriorly and extending to the tip of the abdomen, above which project four pointed processes, which are the tips of the paraprocts and of two accessory structures, the specillum; 10th tergum narrow and behind it a plate-like epiproct, whose shape varies with the species. Terga of some of the other abdominal segments modified, leaving large membranous areas, on the anterior or lateral margins of which,

in some segments, arise chitinised tergal processes, the positions of which form important specific characters.

4. *Capnia* (Fig. 14). Cerci long and many segmented; 9th sternum extending over the 10th; epiproct large and directed forwards towards a chitinised median tergal process on the 8th or 9th tergum. The shape of the epiproct differs in different species.

5. Perlodidae other than *Isoperla* (Fig. 16). Cerci long and many segmented; 10th tergum longitudinally split or entire; if the latter the paraprocts are modified.

6. *Isoperla* (Fig. 17). Cerci long and many segmented; 9th sternum extending back over the tip of the 10th; 8th sternum bearing a heavily chitinised median lobe on its hind margin. In most European species identification is only possible by examination of the internal penial armature (Illies 1952) but the two British species can be satisfactorily separated on the shape of the median lobe (see p. 42).

7. Perlidae (Fig. 18). Cerci long and many segmented; 10th tergum longitudinally divided, each half being modified; epiproct a small plate.

8. *Chloroperla* (Fig. 19). Cerci long and many segmented; 9th sternum extending back over 10th; epiproct bearing a heavily chitinised dorsally directed tooth.

The anatomy of the genitalia and genital system of stoneflies has been discussed in an important paper by Brinck (1956), and terminology for the various parts was agreed upon at the First European Symposium on Plecoptera held in Lausanne in 1956. The terms used above are in conformity with this agreement, and it is to be hoped that they will become universally recognised.

STRUCTURE OF THE NYMPH   The general internal anatomy of the stonefly nymph has been described by Hynes (1941), and Wu (1923) has described the internal and external anatomy of one species in great detail. The general structure is very like that of the cockroach, and the following features are important in the identification of British nymphs.

The head (Fig. 2) bears long antennae the basal segment of which is enlarged, compound eyes and three ocelli (*o*), behind which lies a Y-shaped epicranial suture (*es*), and in front of which is an M-line (*ml*) like that of the adult. Between the lateral ocelli and the bases of the antennae are two scar-like structures, the tentorial callosities (*tc*) the shape of which is sometimes diagnostic. The mouth parts are of the biting type, and points which are used in identification are: the upper lip or labrum (*la*); the 5-segmented maxillary palps (*mp*) the finger-like galea (*g*) and blade-like and often toothed lacinia (*l*) which together terminate the maxilla; the paired inner glossae (*gl*) and outer paraglossae (*pg*), which lie between the three-segmented palps (*lp*) and are the terminal processes of the lower lip or labium; and the submentum (*sm*), the large plate at the base of the labium, which covers most of the lower surface of the head.

The thorax is covered dorsally by three plates the pro-, meso- and meta-nota, laterally by pleura and ventrally by sterna. The

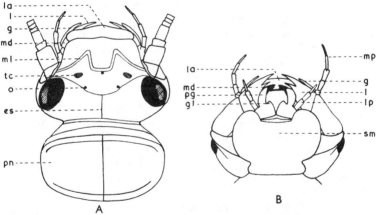

Fig. 2.   A, dorsal view, B, ventral view of head of nymph of *Isoperla grammatica. la,* labrum, *l,* lacinia, *g,* galea, *md,* mandible, *ml,* M-line, *tc,* tentorial callosity, *o,* ocellus, *es,* epicranial suture, *pg,* paraglossa, *gl,* glossa, *mp,* maxillary palp, *lp,* labial palp, *sm,* submentum, *pn,* pronotum, × 14.

developing wing-pads are fused to the meso- and meta-nota, and may differ noticeably in the two sexes of species with short-winged males.

The legs consist of the usual parts, coxa, trochanter, which are both short, long femur and tibia and three-segmented tarsus, which ends in two claws.

The abdomen is ten-segmented and ends in long many-segmented cerci, between which lie two ventral plates, the paraprocts. Each body segment has basically a dorsal tergum and a ventral sternum. These are always separate in the first two segments, and the first sternum is closely associated with the metathorax. The terga and sterna of some or all of segments 3-9 may be wholly or partly fused to one another, forming complete rings. Those of the 10th segment are always so fused, and in the families Taeniopterygidae, Nemouridae, Leuctridae and Capniidae the 10th sternum is very narrow and lightly chitinised, and at first sight appears to be absent. The epiproct, which is so important in adult taxonomy, is fused to the 10th tergum, and in old male nymphs it often appears as a posterior projection. In the Taeniopterygidae all the sterna except the 9th, and sometimes the 8th, are soft and membranous, and in *Brachyptera* and *Rhabdiopteryx* the 9th sternum is drawn out backwards to form a large plate which extends to the tip of the abdomen. In many genera changes of shape of the sterna, the epiproct and the paraprocts, or the first indication of the female aperture on 8th sternum, enable the sexes of older nymphs to be determined. These differences are shown in the figures.

The gills: many species have external gills, which are filiform or sausage-shaped outpushings from the body surface, and which contain tracheae. In British stoneflies these may occur on the prosternum (1st thoracic sternum), pleurae (sides of thoracic segments), coxae, or paraprocts.

Hairs and bristles occur on all species, and are important taxonomically. Three types are distinguished in the keys. Hairs are long thin setae, such as form swimming fringes on the legs; bristles are stouter shorter tapering structures, which are sometimes rounded at the ends and in Nemouridae are often shaped like indian clubs; clothing hairs, which are smaller, may include either of the other types and form a uniform covering over the whole or part of the body. Not only is the occurrence and distribution of these hairs and bristles important, but it is often necessary to know whether they stand more or less upright or lie close against the body. This is best ascertained by examining them in profile against transmitted light.

Coloration and pattern are used as characters in some of the keys, but these should be regarded with caution although they are often useful. This is because in small or newly moulted nymphs the pattern may not have developed, and in nymphs that are ready to emerge adult patterns may show through the nymphal skin.

Ripe nymphs are always recognisable because their wing pads become black, and the wings can be seen crumpled up inside them.

NOMENCLATURE   When this Key was first published (1940) it was stated that synonymy in the order was in confusion. Since 1940 it has been revised for European species, chiefly by Aubert, Brinck and Kimmins. This has, however, resulted in many changes in the names of British species. The following is a list of names which have changed since 1940 together with synonyms and wrongly applied names that have been used in British literature during the past quarter of a century.

Taeniopteryx Pict. = *Nephelopteryx* Klap.
Rhabdiopteryx acuminata Klap. = *R. anglica* Kim. and *R. neglecta* (Alb.)
Brachyptera Newp. = *Taeniopteryx* Klap.
Nemoura cinerea (Retz.) = *N. variegata* Oliv.
Nemoura erratica Claas. = *N. risi* Desp. and *N. marginata* (Pict.)
Amphinemura sulcicollis (Steph.) = *A. cinerea* (Oliv.)
Nemurella picteti Klap. = *N. inconspicua* (Pict.)
Leuctra fusca (L.) = *L. fusciventris* Steph.
Capnia bifrons Newm. = *C. nigra* Pict.
Perlodes microcephala (Pict.) = *P. mortoni* Klap.
Diura bicaudata (L.) = *Dictyopterygella recta* Kemp.
Isoperla Banks = *Chloroperla* Pict.
Isoperla obscura (Zett.) = *I. griseipennis* (Pict.)
Perla bipunctata Pict. = *P. carlukiana* Klap. and *P. marginata* Panz.
Dinocras cephalotes (Curt.) = *Perla cephalotes* Curt.
Chloroperla* Newm. = *Isopteryx* Pict.

The change in *Capnia* and the elimination of the names *Nemoura marginata* and *Perla marginata* are due not to synonymy, but to misidentifications of British species. In *Rhabdiopteryx* British specimens were at first confused with the continental *R. neglecta,* and later it was found that they were the same as a poorly known species from the Baltic area (Kaslauskas 1962).

CLASSIFICATION   The order of the families given here differs from that in the 1940 Key, and there have been some changes in the delimitation of families, even since the recent keys of Kimmins (1950) were published. As compared with the first edition the

* see footnote on p. 45.

Perlidae are restricted, *Chloroperla* and some non-British genera being transferred to a new family, and *Isoperla* being transferred to the Perlodidae. The classification as here set out is that used by Illies (1955) in the most recent large treatise on European stoneflies. This is the classification agreed in 1956 at the First European Symposium on Plecoptera, with the exception that here the various subdivisions of the Nemouridae are regarded as genera, and not as sub-genera. There seems to be little justification for the retention of these groups as sub-genera, although this is done in a large recent comprehensive work on the family (Ricker 1952). All such groupings are merely a matter of opinion or expediency, and in Britain there is no point in retaining the elaboration of sub-genera.

THE BRITISH SPECIES   The British stonefly fauna is of particular interest in that one species is recorded only from Britain, and others are sub-specifically distinct from those on the continent.

Taeniopteryx nebulosa subsp. *britannica* Hynes (1957) differs from the continental form in that males here tend to be short-winged, although not always so, and have epiprocts which are narrower than the cerci instead of wider. The shape of the pronotum of the nymph is also different (Aubert 1950).

Brachyptera putata is apparently confined to Britain, although it is fairly closely related to *B. monilicornis* (Pict.) (Hynes 1957).

Capnia vidua subsp. *anglica* Aubert (1950) differs from the typical form in the male genitalia and the length of the wings of the male.

Isoperla grammatica in Britain belongs, according to Despax (1951), to the sub-species *subarmata* Despax, in which the armature of the male genital atrium (see p. 42) is short. Illies (1952) has, however, shown that the division of this species into subspecies on this character is invalid.

Perlodes microcephala in Britain all belong to the short-winged form, which was until recently called *P. mortoni*. On the continent fully-winged and intermediate specimens occur (Illies 1955).

Perla bipunctata was until recently thought to be specifically distinct in Britain, where it was called *P. carlukiana*, the difference being that British males are always short-winged. Illies (1955) has, however, shown that this appears to be the only difference, and as short-winged males also occur in Morocco (Aubert 1956) it seems doubtful if *carlukiana* should be retained even as a sub-specific name. This requires further investigation.

There are also indications that at least one other British species differs from its continental relatives, and the whole subject of British stoneflies requires further investigation from the point of view of the polytypic species.

One species on the British list may in fact no longer occur here, and another has probably been erroneously recorded. Because there is some doubt about the occurrence of these species here and as any further finds would be of exceptional interest their names are marked with asterisks in the keys.

Isoperla obscura was found early in the century in the river Trent (Morton 1913), where it seems no longer to occur. This species occurs in large rivers on the continent and in Britain it may have been extinguished by pollution, to which stoneflies are very sensitive (Schoenemund 1924). It is, however, possible that the species is still present in some eastern rivers which have been poorly worked for Plecoptera.

Chloroperla apicalis is recorded from Britain because of the existence of three female specimens from Newman's collection now in the Hope Department of Entomology, University Museum, Oxford. As these specimens are without data showing their British origin, Kimmins (1936) has suggested that the record requires confirmation. No specimens have been taken here since, and it seems likely that the record is due to some error. As, however, this species is an inhabitant of large rivers in Europe it may also be one which did once occur here and has become extinct.

KILLING AND PRESERVING Stoneflies preserved in fluid are much easier to identify than dried specimens, which tend to become distorted. The best method of killing is to drop them alive into 70% alchohol; they should then be transferred as soon as they are dead to fresh alcohol or to 2% formaldehyde. It is wise to include 2% glycerin to whichever of these solutions is used as a precaution against drying out. Formalin, although it is unpleasant to work with, has certain advantages over alcohol as a preserving medium; it does not dry up so easily, and it does not bleach the specimens so quickly or completely. It is necessary, however, to wet the specimens thoroughly in alcohol before placing them in formalin, because otherwise they float above the surface, and the formalin does not get at them. When treated as described above it will be found that the specimens float at first just below the surface, but sink after a few days.

The nymphs can be killed and preserved by placing them directly into alcohol or formalin. Here also it is wise to include 2% glycerin, as should the specimen dry out it becomes impregnated with glycerin, does not shrink, and is easily retrievable in fresh solution.

FISHERMEN'S NAMES   Many of the species have been imitated in artificial fly-fishing, and so have received popular names. It is not always easy to fit these to actual species, but this is done here as far as possible. Creepers are the nymphs of Perlidae or the larger Perlodidae. The Stonefly refers to the adults of both species of Perlidae, and presumably to the other large species also. The Yellow Sally is the name given to *Isoperla grammatica*. The February Red is the female of *Taeniopteryx nebulosa,* and the Early Brown is the name applied to species of the family Nemouridae which emerge early in the year, including the common *Protonemura meyeri*. The Willow Fly is *Leuctra geniculata,* and the other species of *Leuctra* are called Needle Flies. Other species in the order do not seem to have received popular names.

ECOLOGICAL NOTES   After the name of each species in the keys to the adults the following information is given: the length of the body of the male and female in mm, the months during which adults have been collected in Britain, an indication of the commonness and abundance of the species, and notes on the habitats of the nymphs. In the keys to the nymphs the page on which these notes occur and the range of length of the fully grown nymphs are given after each specific name. In the notes on commonness and abundance the words 'common', 'rare' or 'local' indicate whether the species occurs in many or few localities in any one part of the country or whether it is confined to one area, while the words 'abundant' or 'frequent' refer to the actual numbers in any one locality. The word 'frequent' is used rather than 'scarce' as in No. 16 of this series because where stonefly species occur at all it is always possible to find several even of the scarcest species. Abundant species can often be collected in great numbers.

A few normally fully-winged species tend to become short-winged at high altitudes (Kühtreiber 1934, Hynes 1941). This is quite distinct from the short-wingedness of the males of many species as it involves both sexes. Where this is known to occur it has been noted after the specific name. The data on which these notes are based has been amassed during many years' study of the insects in various parts of the country, from the literature and from collections in the British Museum and elsewhere. Much information both on the ecology and distribution of the species has been obtained from collections sent to me for identification.

Adult stoneflies live for only a few days and are rarely found far from water, as even those which fly well are not sustained fliers. The habitats of the nymphs are therefore more important than those of the adults in the ecology of the order. Adults of the families Perlodidae, Perlidae and Chloroperlidae do not feed at all, although they drink water freely. Adults of the other families scrape lichen and algae off trees and fences, and it has been shown that at least for *Nemoura cinerea* and *Capnia bifrons* this is necessary to allow them to live long enough to produce eggs (Hynes 1942, Khoo 1964).

Mating occurs on the ground very soon after the emergence of the females, which usually begin to appear a few days after the first males. A few days after mating the eggs are extruded and carried by the females on the under side of the tip of the abdomen in a mass held together with a sticky substance. The eggs are then deposited on the water surface. In the larger species this is done by swimming or running on the water, where the sticky substance dissolves and the eggs fall apart and sink. The smaller species fly down and dip the egg-mass into the water, where it rapidly disintegrates. Each female apparently produces two or three masses of eggs. The eggs are provided with attachment mechanisms which anchor them to the bottom (Percival & Whitehead 1928, Hynes 1941).

The fully ripe nymphs, which are easily recognisable because of their black wing-pads, crawl out of the water, usually at night, and find sheltered places under stones etc. on the bank, in which to emerge. The adults remain in such places, running actively about and flying to nearby trees and bushes only when the air is calm and warm.

Most British species take a year to develop from egg to adult, but *Perla* and *Dinocras* take three years. It is possible that *Chloroperla tripunctata* takes two years. The eggs of many species take many weeks to hatch (Hynes 1962) and the small nymphs are difficult to find without special techniques. The nymphs of *Capnia bifrons* also disappear from collections when they go into diapause at an early stage in their growth (Khoo 1964). There are therefore periods of three or four months in each year during which many species appear to be absent from localities where they are actually abundant. For those emerging in the early months of the year, which group includes most of our species, this period coincides with "summer holidays". Once hatched the nymphs grow steadily even at low winter temperatures. Those of the families Perlidae, Perlodidae and Chloroperlidae are carnivorous, but all also eat some vegetation. Nymphs of the other families are herbivorous.

Stonefly habitats fall roughly into five classes: (1) still waters (ponds, tarns and dykes) with emergent vegetation; (2) the stony shores of lakes; (3) the emergent vegetation on the banks of rivers and streams; (4) small streams with stony beds; (5) rivers with stony substrata and moss on the larger stones.

GEOGRAPHICAL DISTRIBUTION   The geopraphical distribution of stoneflies in continental Europe has been the subject of several papers in recent years (Illies 1953, 1955, Rauŝer 1962), and they are clearly a group of considerable significance in historical limnology.   On pp. 82-87 maps are given showing the known distribution of the various species in Britain.   These maps are based on the vice-county system (Balfour-Browne 1931) and the whole vice-county is shaded if the species is certainly known to occur there.   This information is clearly incomplete, and the co-operation of all collectors is sought to help to complete it.   At present only the Lake District, Yorkshire, the Isle of Man and some parts of North Wales have been thoroughly studied, but even in these areas there may still be unrecorded species.   The present range of the species if known completely would give much information on their ecology and past history.

For some species there is too little information to warrant the production of a map, and for others there is some doubt about certain records, which have not been included in the maps.   These species are:

Rhabdiopteryx acuminata has been found only in north-east and mid-west Yorkshire, Radnorshire and Montgomeryshire.

Brachyptera putata (map p. 83) was only doubtfully recorded from Herefordshire in the last edition of this booklet.   Its continued occurrence there has since been confirmed.

Leuctra nigra (map p. 85) Hynes (1941) records the nymph of this species from Co. Dublin.   This was a misidentification.

Capnia atra (map p. 86) Morton (1929) records this species from a lake in Co. Kerry.   He states, however, that the males were short-winged, but as no other British populations with short-winged males are known, and where the species is short-winged in Sweden both sexes are involved, this record needs confirmation. This is especially so as this is the only British record outside Scotland.

Isogenus nubecula remained for many years on the British list because of the existence of a few unlabelled specimens in various old collections. Recently the species has been found in the River Dee where it forms the boundary between the detached bit of Flintshire, which forms part of Cheshire in the vice-county system, and Denbighshire (Hynes 1963). There is also an old, unconfirmed record from near Stoke-on-Trent, Staffordshire (Morton 1934), which may be the unlabelled specimen in the McLachlan collection at the British Museum.

Diura bicaudata. A female specimen in the Dale collection is labelled "Charmouth" (Kimmins 1944), but, as there are no other records from south-western England, this needs confirmation.

Isoperla obscura is recorded only from the river Trent near Nottingham, where it has been found on three occasions early in the century. No trace of it could be found there in 1940.

Chloroperla apicalis. The three allegedly British specimens are without locality labels.

ACKNOWLEDGEMENTS   My thanks are due to Mr D. E. Kimmins for much helpful advice and information, and to Drs Aubert, Brinck and Illies for gifts and loans of specimens from the Continent. I am grateful also to the Council of the Royal Entomological Society for permission to reproduce parts of figures from my papers published by them (Hynes 1941, 1955, 1957 and 1963): these are in Figs. 20, 21, 22, 24, 25, 26, 28, 30, 31, 32, 33, 35, 36, 37, 38, 39, 41, 42, 43, 44, 45, 46 and 47. Many people, too numerous to mention here, have also helped me in the preparation of the distribution maps by sending me specimens to identify. To them all I express my grateful thanks.

THE ADULTS

KEY TO FAMILIES

1   Cerci short, not longer than the greatest width of the pronotum—   **2**

—   Cerci long   **4**

2   All three tarsal segments about equally long (Fig. 3A)— TAENIOPTERYGIDAE, p. 23

—   Second tarsal segment shorter than first or third (Fig. 3B)—   **3**

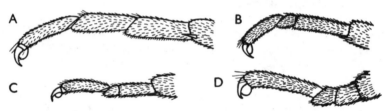

Fig. 3.   Hind tarsi of adults, × 30: A, *Rhabdiopteryx acuminata*; B, *Protonemura praecox*; C, *Capnia bifrons*; D, *Isoperla grammatica*.

3   In both wings the veins Sc 1, Sc 2, R 4 + 5 and the cross-vein *r-m* form a distinct X-like figure (Fig. 4A). Wings at rest held flat over the abdomen—   NEMOURIDAE, p. 27

—   No such X-like formation of the veins (Fig. 4B). Wings at rest rolled round the abdomen—   LEUCTRIDAE, p. 33

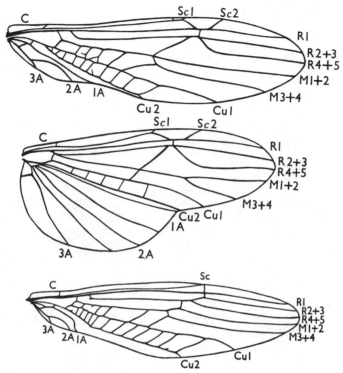

Fig. 4. Right wings, × 10: A, fore and hind-wings of *Nemurella picteti*; B, fore-wing of *Leuctra fusca*.

4(1)* Second tarsal segment much shorter than the first or third, which are sub-equal (Fig. 3C). Small dark-coloured species; males often with very short wings— CAPNIIDAE, p. 37

— Third tarsal segment longer than the first and second together (Fig. 3D). Large dark-coloured species, or smaller green species— 5

5 Short-winged males— 6

— Fully winged specimens— 7

* Where a couplet is not reached directly from the preceding one, the number of the couplet from which the direction came is indicated thus in parentheses.

6   10th abdominal tergum longitudinally divided, each half
    bearing a forwardly directed process on its inner margin
    (Fig. 18)—   PERLIDAE, p. 43

—   10th abdominal tergum entire (Figs. 16c and E)—
                                            PERLODIDAE, p. 39

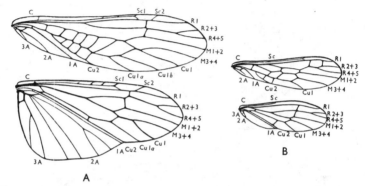

Fig. 5.   Right wings, × 7 : A, *Isoperla grammatica*; B, *Chloroperla
          torrentium*.

7(5) Anal area of hind wing small; all anal veins simple (Fig. 5B)—
                                            CHLOROPERLIDAE, p. 45

—   Anal area of hind wing large; 2A and 3A branched (Figs. 1,
    5A)—   8

8 In the fore wing cross-vein *r-m* arises from R4 + 5 (Fig. 1A). Large dark species with no median longitudinal yellow line on the pronotum— PERLIDAE, p. 43

— In the fore wing cross-vein *r-m* arises from Rs before it divides into R2 + 3 and R4 + 5 (Figs. 5A and 6). Green species or large dark species with a clear longitudinal yellow line on the pronotum— PERLODIDAE, p. 39

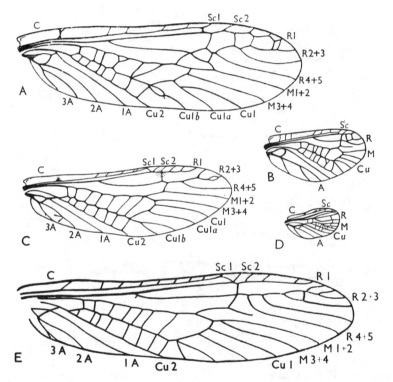

Fig. 6. Right fore-wings, × 5: A, female *Perlodes microcephala*; B, male *P. microcephala*; C, female *Diura bicaudata*; D, male *D. bicaudata*; E, female *Isogenus nubecula* (drawn from a German specimen).

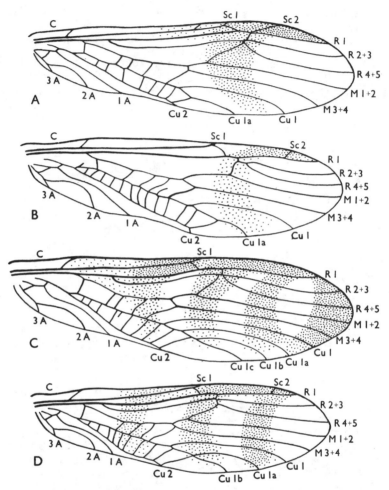

Fig. 7. Right fore-wings, × 9: A, female *Rhabdiopteryx anglica*;
B, female *Taeniopteryx nebulosa*; C, female *Brachyptera
putata*; D, female *B. risi*.

Family TAENIOPTERYGIDAE Klapálek

(See Illies 1955, Hynes 1957)

1 Fore-wing with two branches to the vein Cu 1 (Figs. 7A and B) and with only one dark crescentic transverse band, which is often indistinct— **2**

— Fore-wing with more than two branches to the vein Cu 1 and with 3 dark bands (Figs. 7C and D).
(*Taeniopteryx* Klap.) genus BRACHYPTERA Newport **3**

2 Each coxa with a scar on the inner posterior side (Fig. 8A). Fore-wing with no cross-veins between C and Sc (Fig. 7B). Males often short-winged, with two-segmented cerci (Fig. 8C), and a drumming lobe on the 9th sternum (Fig. 8B). Female sub-genital plate (9th sternum) broad and little drawn out posteriorly (Fig. 8C)—
(*Nephelopteryx* Klap.) genus TAENIOPTERYX Pictet

One British species **Taeniopteryx nebulosa** (Linn.) (Aubert 1950).

Male 7-9; female 9-11 mm. Feb.-May, mainly Feb. to Mar. Fairly rare, frequent. Emergent vegetation of rivers, occasionally in moss, but usually in sedge, reeds or grass.

— Coxae without scars. Fore-wing usually with one or two cross-veins between C and Sc (Fig. 7A). Males fully winged, with cerci of several segments (Fig. 8D) and without a drumming lobe on the 9th sternum. Female sub-genital plate long and drawn out to a point (Fig. 8D)—
genus RHABDIOPTERYX Klapálek

One British species **Rhabdiopteryx acuminata** Klapálek (R. *anglica* Kim), confused in the past with R. *neglecta* (Alb.) (Kimmins 1943).

Male 8-9; female 8-10 mm. Mar.-May. Local, apparently abundant. Small calcareous rivers in Yorkshire; recorded from Radnorshire and Montgomeryshire.

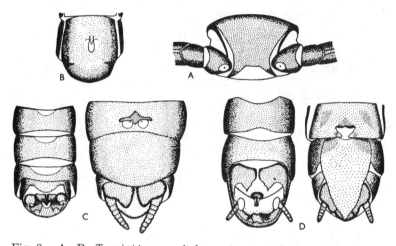

Fig. 8.   A, B, *Taeniopteryx nebulosa*: A, ventral view of prothorax
and bases of legs of male, × 18; B, ventral view of tip
of abdomen of male, × 18.
C, D, (on left) male genitalia from above, (on right) female
from below, × 18: A, *Taeniopteryx nebulosa*; B, *Rhabdiopteryx acuminata*.

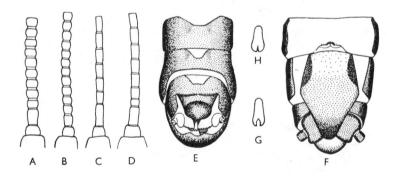

Fig. 9.   A-D, basal few segments of antennae, × 20: A, male
Brachyptera putata; B, female *B. putata*; C, male *B. risi*;
D, female *B. risi*.
E, F, genitalia × 20; E, male *B. putata* in dorsal view;
F, female *B. risi* in ventral view.
G, H, tips of male epiprocts × 60; G, *B. putata*; H, *B. risi*.

3(1) Antennae moniliform (like a string of beads), the basal segments being as wide as long (Figs. 9A and B). Apex of fore-wing darkened (Fig. 7c). Male short-winged. Female with 4 or more branches to Cu 1 in fore-wing (Fig. 7c)—

Brachyptera putata (Newman)
(Morton 1911)

Male 7-9; female 8-10 mm. Mar.-Apr. and probably Feb. Rare, apparently frequent. Slower reaches of Scottish rivers; and in Herefordshire.

— Antennae filiform (parallel-sided), the basal segments being longer than wide (Figs. 9c and D). Apex of fore-wing not darkened (Fig. 7D). Male fully winged. Female with 3, rarely 4, branches to Cu 1 in fore wing (Fig. 7D)—

Brachyptera risi (Morton)

Male 7-10; female 8-11 mm. Mar.-July, occasionally Feb., mainly Mar.-May. Common, fairly abundant. Small stony streams, occasionally in moss in rivers.

The differences between the genitalia of these two species are small, but those of a male *B. putata* and a female *B. risi* are shown in Fig. 9 to demonstrate the difference between the sexes. In the males the shape of the upturned edge of the 9th sternum differs, being lightly emarginate in *B. putata* and broadly pointed in *B. risi*; the drumming lobe of *B. putata* is much smaller than that of *B. risi*; and the shape of the tip of the epiproct is slightly different in the two species as shown in Figs. 9G and H. In the females the shapes of the various parts are almost identical, but in *B. risi* there is normally a well-defined dark patch on each side of the sub-genital plate, which is absent from *B. putata*.

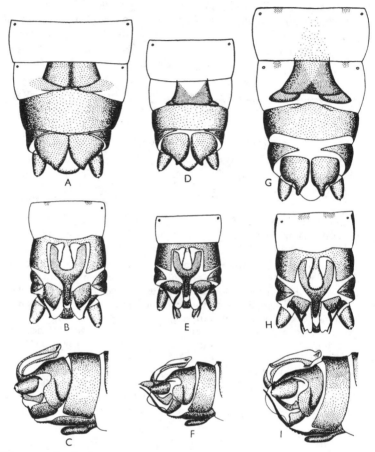

Fig. 10.   Genitalia, × 25 : top row, females in ventral view; middle
row, males in ventral view; bottom row males in lateral
view.
A, B, C, *Protonemura praecox*; D, E, F, *P. montana*;
G, H, I, *P. meyeri*.

Family NEMOURIDAE Klapálek
(See Kimmins 1940, Despax 1951, Illies 1955)

1 Vestiges of nymphal gills present on the prosternum. Cerci of male simple and acorn-shaped (Figs. 10 and 11). 8th sternum of female bearing a median chitinised plate (Figs. 10A, D, G and 11A) or with an excised hind margin (Fig. 11B)—   **2**

— No vestiges of nymphal gills on prosternum. Cerci of male modified into copulatory organs (Figs. 12A, C, D, and E) or simple and cylindrical (Figs. 12B and F). 8th sternum of female unmodified (Figs. 12A, B, C and D), or bearing a rounded protuberance on each side of the sub-genital plate (Fig. 12F), or with three shallow longitudinal grooves anteriorly (Fig. 12E)—   **6**

2 Three sausage-shaped prosternal gill-vestiges on each side (Fig. 26D). External appendages of the male paraprocts long, separate from the sub-anal plates and with lateral armature (Fig. 10). 8th sternum of female with a median chitinised plate and two postero-lateral plates (Fig. 10)—
                genus PROTONEMURA Kempny   **3**

— Two bunches of 5 to 8 filamentous prosternal gill-vestiges on each side (Fig. 28D). External appendages of the male paraprocts short, fused to the sub-anal plates and without armature (Fig. 11). 8th sternum of female with only a median chitinised plate (Fig. 11A), or with a three-lobed excision posteriorly (Fig. 11B)—   genus AMPHINEMURA Ris   **5**

3 Sub-anal plates of male paraprocts with a long posteriorly directed spine-like process, which is longer than the width of the plate (Figs. 10E, F, H and I). Posterior margin of the median plate of the 8th sternum of female a straight line or slightly concave (Fig. 10D and G)—   **4**

— Sub-anal plates of male with a very short posterior process (Figs. 10B and C). Posterior margin of the female median plate widely convex (Fig. 10A)—
                Protonemura praecox (Morton)

Male 6-8, female 7-9 mm. Feb.-May, occasionally June, mainly Feb.-Apr. Fairly rare, frequent. Small swift stony streams up to high altitudes.

4   Apex of male external appendage bent at right angles, so that
    it is directed tailwards (Fig. 10F). Postero-lateral plates of the
    8th sternum of female exposed only beside the hind corners
    of the median plate, the corners of which are truncate; female
    paraprocts less drawn out posteriorly (Fig. 10D). Head
    uniformly dark brown—   **Protonemura montana** Kimmins
                                                    (Kimmins 1941)
    Male 6-8; female 7-9 mm. Aug.-Sept. Rare, frequent. Small
    stony streams at altitudes above about 1700 ft.

—   Apex of male external appendage not bent at right angles
    (Fig. 10 I). Postero-lateral plates of 8th sternum of female
    exposed beside and behind the median plate, the corners of
    which are pointed; female paraprocts more drawn out
    posteriorly (Fig. 10G). Head usually with a transverse pale
    bar across the top—   **Protonemura meyeri** (Pictet)
    Male 5-8; female 7-9 mm. Feb.-June, occasionally as late as
    Sept., mainly Mar.-May. Very common, very abundant. Swiftly
    flowing water, especially where aquatic mosses occur; up to high
    altitudes.

    Males of *Protonemura* are also easily identified by the shape of
    the epiproct as seen in side view (Figs. 10C, F and I).

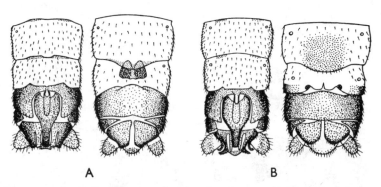

A   **B**

Fig. 11. Genitalia (male on left, female on right) from below, × 37·5: A, *Amphinemura sulcicollis*; B, *A. standfussi*.

5(2) External appendage of male paraproct bifid with hairs at the tip, the outermost part bent round the base of the cercus (Fig. 11B). Male epiproct spatulate in side view with a few spines on its lower surface. Posterior margin of 8th sternum of female with a three-lobed excision (Fig. 11B)—
Amphinemura standfussi Ris
Male 4-5; female 5-6 mm. June-Sept. Rare, frequent. Small stony streams: apparently associated with large amounts of vegetable matter.

— External appendage of male simple, without hairs at the tip, and bent round the base of the cercus (Fig. 11A). Male epiproct with a beak-like projection at the tip and without spines on its lower surface. 8th sternum of female with a median chitinised plate, which may appear divided down the centre, and at most a small median notch posteriorly (Fig. 11A)—
(*A. cinerea* (Oliv.))   **Amphinemura sulcicollis** (Stephens)
Male 4-6; female 5-7 mm. Apr.-Sept., mainly Apr.-June. Very common, very abundant. Running water with a stony sub- stratum; usually in larger streams and rivers.

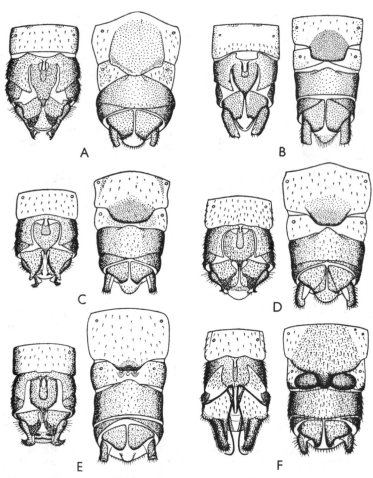

Fig. 12.   Genitalia (male on left, female on right) from below × 25:
A, *Nemoura cinerea*; B, *N. dubitans*; C, *N. cambrica*;
D, *N. erratica*; E, *N. avicularis*; F, *Nemurella picteti*.

6(1) Male cerci usually modified; paraprocts undivided and without long processes (Figs. 12A-E). Female with 8th sternum unmodified (Figs. 12A-D) or with three shallow longitudinal grooves anteriorly (Fig. 12E)—   genus NEMOURA Pictet **7**

— Male cerci simple and cylindrical; paraprocts sub-divided, with a long filamentous process on the sub-anal plate reaching at least to the tip of the cerci (Fig. 12F). Female with a large hemispherical protuberance on the 8th sternum on each side of the narrow sub-genital plate (Fig. 12F)—
genus NEMURELLA Kempny

One British species **Nemurella picteti** Klapálek (*N. inconspicua* (Pict.)).

Male 6-7; female 6-9 mm. Feb.-Sept. Common, abundant. Still or slow-flowing water with much emergent vegetation, and small mossy trickles particularly at high altitudes.

7 Male cerci with two distinct chitinised points, being shaped like a pick-axe whose points are directed dorsally and ventrally (Fig. 12A). Female sub-genital plate (7th sternum) 3-lobed and extending to 9th sternum (Fig. 12A). Pronotum dull and coarsely punctate—
(*N. variegata* Oliv). **Nemoura cinerea** (Retzius)

Male 6-7; female 6-9 mm. Mar.-Sept., mainly Mar.-July. Very common, very abundant. Still or slow-flowing water with much emergent vegetation; occasionally in sluggish stony streams.

— Male cerci with only one or no chitinised points. Female sub-genital plate not 3-lobed and shorter. Pronotum shiny, not coarsely punctate—   **8**

8 Male cerci not hooked at apices (Fig. 12B). Female sub-genital plate large, truncate at the end, and occupying more than half the width of the 7th sternum (Fig. 12B)—

Nemoura dubitans Morton

Male 5-6; female 6-8 mm. Apr. Rare, probably frequent. Nymphal habitat small shallow overgrown spring streams.

— Male cerci hooked at apices. Female sub-genital plate widely rounded, or if truncate occupying less than half the width of the 7th sternum— **9**

9 Tip of male cercus half-moon-shaped with a distinct remnant of the 2nd segment at the tip (Fig. 12E). Female sub-genital plate small and truncate, over-lying (often completely concealing) a 3-lobed projection, the tips of which lie in grooves on the 8th sternum (Fig. 12E)—

Nemoura avicularis Morton

Male 6-8; female 7-9 mm. Mar.-Aug., mainly Apr.-May. Fairly common, abundant. Lake shores and occasionally in emergent vegetation of rivers.

— Tip of male cercus more or less foot-shaped and without vestige of the second segment. Female sub-genital plate simple, not overlying a 3-lobed projection— **10**

10 Male paraprocts produced backwards into short finger-like processes, cerci with small hairs on the hind margin of the laterally directed "toe" (Fig. 12C). Female sub-genital plate wide and rounded, and occupying more than half the width of the 7th sternum (Fig. 12C)—

Nemoura cambrica (Stephens)

Male 5-6; female 6-8 mm. Apr.-June. Fairly common, abundant. Small stony streams, particularly where many leaves are present.

— Male paraprocts rounded at the tip and markedly emarginate on the inner side, cerci without hairs on the laterally directed "toe", which is a simple chitinised point (Fig. 12D). Female sub-genital plate truncate and occupying less than half the width of the 7th sternum (Fig. 12D)—

(*N. risi* Desp.) **Nemoura erratica** Claassen

(Confused in Britain with *N. marginata* (Pict.)).

Male 5-6; female 5-7 mm. Feb.-Sept., mainly Mar.-June. Rare, fairly abundant. Small stony streams. Tends to be short-winged at high altitudes.

Family LEUCTRIDAE Klapálek
Genus LEUCTRA Stephens

(See Mosely 1932, Despax 1951, Illies 1955)

1   Antennae with a whorl of outstanding hairs round the apex of each segment. In the male a single median process on 8th abdominal tergum and two small processes close together on 7th tergum (Fig. 13A). Female sub-genital plate (8th sternum) not divided at the apex into two distinct lobes (Fig. 13A)—   **Leuctra geniculata** (Stephens)

Male 7-9; female 8-11 mm. Aug.-Nov., mainly Aug.-Sept.; single records in Apr. and May. Fairly common, abundant. Stony beds of large streams and rivers.

—   Antennae without a whorl of outstanding hairs at the apex of each segment. In the male all tergal processes paired or absent. Female sub-genital plate divided posteriorly into two distinct lobes—   **2**

2   Male without tergal processes on any segment (Fig. 13B). Female with posterior margin of each lobe of sub-genital plate sinuate (Fig. 13B)—   **Leuctra inermis** Kempny

Male 4-6; female 5-7 mm. Apr.-Aug., mainly Apr.-June; single record in Feb. Very common, very abundant. Rivers and streams with stony substrata. Tends to be short-winged at high altitudes.

—   Male with paired tergal processes on at least one abdominal segment. Female with posterior margins of lobes of sub-genital plate not sinuate—   **3**

3   Male with tergal processes only on 8th segment (Fig. 13D). Female with each lobe of sub-genital plate drawn out backwards and inwards leaving a Y-shaped area between the lobes (Fig. 13D)—   **Leuctra hippopus** (Kempny)

Male 5-7; female 6-9 mm. Feb.-June, mainly Feb.-Apr. Common, abundant. Rivers and streams with a stony substrata. Tends to be short-winged at high altitudes.

—   Male with tergal processes on two segments. Female with lobes of sub-genital plate not so produced—   **4**

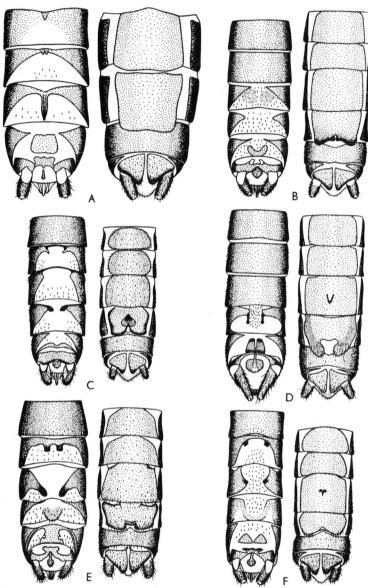

Fig. 13. *Leuctra,* genitalia (male from above on left, female from below on right) × 25:
A, *L. geniculata*; B, *L. inermis*; C, *L. nigra*; D, *L. hippopus*; E, *L. fusca*; F, *L. moselyi*.

4   Male with tergal processes on 6th and 8th segments (Fig. 13C). Female sub-genital plate with a small central lobe between the two main lobes and a heavily chitinised area in the middle (Fig. 13C)—   **Leuctra nigra** (Olivier)

Male 4-6; female 5-8 mm. Apr.-Aug., mainly Apr.-May. Fairly common, very abundant. Small stony streams, particularly in very silty reaches. Tends to be short-winged at high altitudes.

—   Male with tergal processes on 6th and 7th segments. Female sub-genital plate without a central lobe or heavily chitinised area—   **5**

5   Male with tergal processes on 6th segment close together on the anterior margin of the membranous area and pointing towards the tail (Fig. 13E). Female with truncated lobes on sub-genital plate (13E)—   **Leuctra fusca** (Linné) (*L. fusciventris* Steph.)

Male 6-8; female 7-9 mm. June-Dec., mainly Aug.-Oct. Very common, very abundant. All types of water with a stony substratum.

—   Male with tergal processes on 6th segment at the sides of the membranous area and pointed slightly inwards (Fig. 13F). Female with widely rounded lobes on sub-genital plate (Fig. 13F)—   **Leuctra moselyi** Morton

Male 5-6; female 6-7 mm. July-Aug. Rare, abundant. Small stony streams.

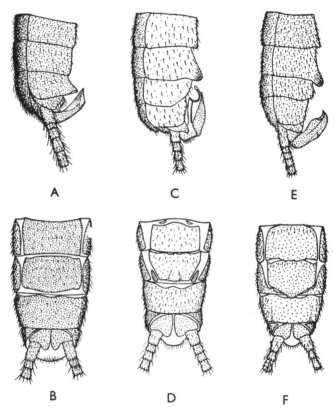

Fig. 14. *Capnia,* genitalia (upper row male from the left side, lower female from below) × 25.
A, B, *C. bifrons;* C, D, *C. atra;* E, F, *C. vidua.*

Family CAPNIIDAE Klapálek
Genus CAPNIA Pictet
(See Despax 1951, Illies 1956)

1 Male short-winged, 9th tergum drawn out into an upwardly directed cone, epiproct simply pointed at the tip (Fig. 14A). Female sub-genital plate (8th sternum) not produced apically and without lateral chitinised plates (Fig. 14B). Between veins Cu 1 and Cu 2 in both wings a quadrilateral cell narrowing towards the base (Fig. 15A)   **Capnia bifrons** (Newman) (Confused in the past with *C. nigra* Pict.)

Male 4-6; female 5-8 mm. Feb.-May, mainly Mar.-Apr. Fairly rare, abundant. Small stony streams and stony lake shores.

— Male short or fully winged, 8th tergum drawn out into an upwardly directed spinose knob, epiproct distinctly emarginate on the under side. Female sub-genital plate produced into a triangle at the apex or with lateral chitinised plates. Between Cu 1 and Cu 2 in both wings a triangular cell (Fig. 15B)— 2

2 Male fully winged, epiproct very emarginate on the under side so that a long thin projection extends forwards (Fig. 14C). Female sub-genital plate not produced posteriorly and with a small heavily chitinised plate on each side (Fig. 14D)— **Capnia atra** Morton

Male 5-6; female 5-7 mm. Mar.-May, probably Feb. also. Fairly common in Scotland, abundant. Stony lake shores, occasionally in stony lake outflows.

— Male short winged, epiproct slightly emarginate on the under side (Fig. 14E). Female sub-genital plate produced posteriorly into a blunt triangle and without lateral plates (Fig. 14F)— **Capnia vidua** Klapálek (Aubert 1950)

Male 4-6; female 5-8 mm. Mar.-Apr. Rare, often scarce. Small stony streams.

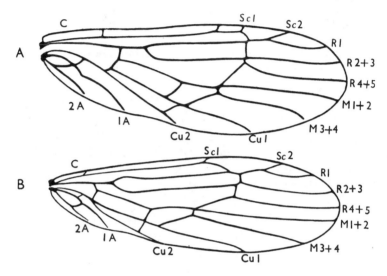

Fig. 15. *Capnia,* right fore-wings, × 20 : A, female *C. bifrons*; B, female *C. vidua.*

Family PERLODIDAE Klapálek

(See Despax 1951, Illies 1955)

1 Large dark-coloured species with a distinct yellow line down the middle of the pronotum. Males may be short-winged. In fore-wing R 2 + 3 at least 2-branched (Fig. 6)— **2**

— Medium sized green species. Males never short-winged. In the fore-wing R 2 +3 simple (Fig. 5A)— (*Chloroperla* Pict.) genus ISOPERLA Banks **4**

2 Male fully winged, tenth tergum longitudinally divided (Fig. 16A). Female sub-genital plate (posterior projection of 8th sternum) occupying nearly the whole width of 9th sternum (Fig. 16B). In fore-wing cross-veins *r* and *r-m* separated by less than one-third the length of each (Fig. 6E)—
ISOGENUS Newman

One British species **Isogenus nubecula** Newman

Male 14-19; female 15-20 mm. Mar.-Apr. Local, frequent. Large stony lowland rivers; recorded recently only from the Welsh Dee.

— Male short-winged, tenth tergum undivided. Female sub-genital plate occupying little more than three-quarters or less of the width of the 9th sternum. In fore-wing (of females) cross-veins *r* and *r-m* separated by half or more of the length of each (Figs. 6A and c)— **3**

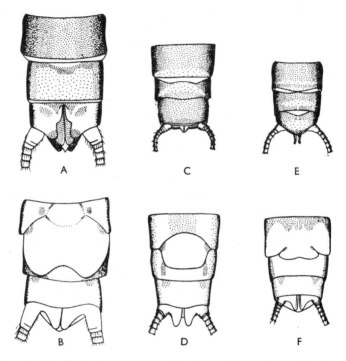

Fig. 16. Genitalia (upper row male from above, lower female from below) × 12·5:

A, B, *Isogenus nubecula*; C, D, *Perlodes microcephala*; E, F, *Diura bicaudata*.

(A and B drawn from German specimens).

3   Tip of fore-wing with an irregular network of cross-veins
    (Figs. 6A and B). Male paraprocts short and triangular, each
    extending beyond 10th tergum by less than half its length
    (Fig. 16c). Female sub-genital plate occupying just over
    three-quarters of the width of the 9th sternum (Fig. 16D)—
                                            genus PERLODES Banks
        One British species **Perlodes microcephala** (Pictet) (*P.
    mortoni* Klap.).

    Male 13-18; female 16-23 mm. Mar.-July, mainly Mar.-May.
    Common, frequent. Rivers and streams with stony substrata
    up to about 1200 ft.; occasionally on stony lake shores in Scot-
    land

—   Tip of fore-wing without an irregular network of cross-veins
    (Figs. 6c and D) except in a few males where it is always less
    developed than in *Perlodes*. Male paraprocts elongate, each
    shaped like a half-cylinder and extending beyond the 10th
    tergum by more than half its length (Fig. 16E). Female sub-
    genital plate occupying about two-thirds of the width of the
    9th sternum (Fig. 16F)—   genus DIURA Billberg
        One British species **Diura bicaudata** (Linné) (*Dictyoptery-
    gella recta* Kemp; see Brinck 1949).

    Male 10-13; female 12-14 mm. Apr.-June, one record in Aug.
    Common, often abundant. Stony lake shores, stony streams at
    altitudes above 1,000 ft.

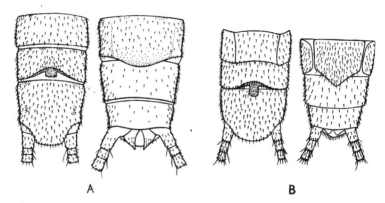

Fig. 17. Male (left) and female (right) genitalia from below, × 15:
A, *Isoperla grammatica*; B, *I. obscura*.

4(1) Chitinised lobe on hind .margin of 8th sternum of male rectangular and twice as wide as long (Fig. 17A). Female sub-genital plate wide and rounded (Fig. 17A). Femora of all legs with longitudinal black stripes—
Isoperla grammatica (Poda)

Male 8-11; female 9-13 mm. Apr.-Aug. Very common, very abundant. Stony rivers and streams; stony lake shores in the far north.

— Chitinised lobe on hind margin of 8th sternum of male slightly longer than wide and rounded at the tip (Fig. 17B). Female sub-genital plate triangular (Fig. 17B). Femora without black stripes— **Isoperla obscura*** (Zetterstedt). (*I. griseipennis* (Pict.)).

Male 8-10; female 9-12 mm. May-June. Local, possibly extinct. R. Trent near Nottingham. On the continent in large lowland rivers, where it is always scarce.

On the continent it is necessary to examine the armature of the male genital cavity to be certain of specific identity (see Illies 1952 and 1955 for technique). This armature consists of an area of overlapping teeth which can be seen by transparency through the 9th sternum after maceration. In *I. grammatica* the area is an elongate strip, and in *I. obscura* it is V-shaped with the apex posteriorly.

* see p. 13

Family PERLIDAE McLachlan
(See Illies 1955)

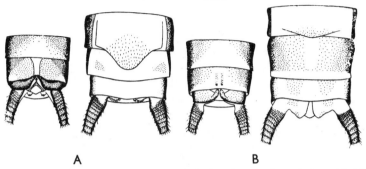

A   B

Fig. 18.   Male genitalia from above (left), female from below (right),
× 12·5 : A, *Dinocras cephalotes*; B, *Perla bipunctata*.

1   In the hind-wing one to three cross-veins (which are often very
weak) in the cell bounded by M, Cu and *m-cu* (Fig. 1B).   In
the male each half of the longitudinally cleft 10th tergum gives
rise to a forwardly pointing projection whose base occupies
the entire inner margin of the plate (Fig. 18A).   Female sub-
genital plate projecting far over 9th sternum (Fig. 18A).
Pronotum and area in front of M-line black—
genus DINOCRAS Klapálek

One British species **Dinocras cephalotes** (Curtis) (*Perla
cephalotes* Curt.).

Male 15-19; female 18-24 mm. Apr.-June, mainly May-June.
Fairly common, fairly abundant.   Rivers with stable rocky and
stony substrata, occasionally in streams.

—   In the hind wing no cross-veins in the cell bounded by M,
Cu and *m-cu* (Fig. 1A).   In the male the forwardly pointing
projection of the longitudinally cleft 10th tergum arises from
the hinder half of the inner margin of the plate (Fig. 18B).
Female sub-genital plate short (Fig. 18B).   Pronotum pale
yellow, with a black border and median longitudinal line on
each side of which lies a dark patch (Fig. 1A).   Area in front
of M-line yellow and grey at the sides— genus PERLA Geoffroy

One British species **Perla bipunctata** Pictet (*P. carlukiana*
Klap., also misidentified in Britain as *P. marginata* Panz.).

Male 16-20; female 18-24 mm. Apr.-June, mainly May-June.
Common, abundant.   Rivers and streams with unstable stony
substrata.

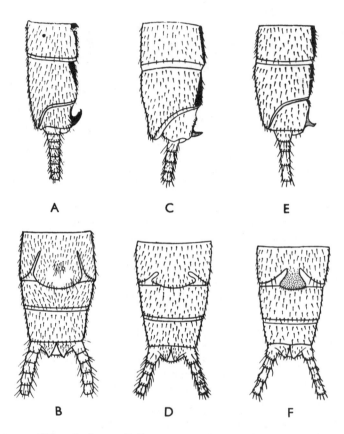

Fig. 19.   *Chloroperla*, genitalia upper row male from the left, lower female from below, × 25:
A, B, *C. torrentium*; C, D, *C. tripunctata*; E, F, *C. apicalis*;
E from a French and F from a Corsican specimen.

Family CHLOROPERLIDAE Okamoto

Genus CHLOROPERLA† Newman (*Isopteryx* Pictet)

(See Kimmins 1936, Despax 1951, Illies 1955)

1 Tooth of male epiproct heavily chitinised and emarginate below the tip (Fig. 19A). Female sub-genital plate more than half as long as 8th sternum and with a conspicuous bunch of hairs in the centre (Fig. 19B)— **Chloroperla torrentium**† (Pictet)
Male 5-7; female 6-8 mm. Apr.-Aug., mainly Apr.-June. Very common, very abundant. All types of water with a stony substratum.

— Tooth of male epiproct not emarginate below the tip. Female sub-genital plate shorter and without a bunch of hairs in the centre— **2**

2 Front margin of male epiproct tooth straight and more heavily chitinised than the rest of the tooth (Fig. 19C). Female sub-genital plate no more heavily chitinised than the sterna, and two-thirds the width and less than one-half the length of the 8th sternum (Fig. 19D)— **Chloroperla tripunctata** (Scopoli)
Male 6-7; female 7-9 mm. May-July, mainly May-June. Fairly common, fairly abundant. Rivers and streams with stony substrata.

— Front margin of male epiproct tooth emarginate and only slightly more heavily chitinised than the rest of the tooth (Fig. 19E). Female sub-genital plate more heavily chitinised than the sterna, and one-third the width and one-half the length of the 8th sternum (Fig. 19F)—
Chloroperla apicalis*† Newman
Male 5.5-7; female 6-8 mm. On the continent June-July. Large lowland rivers. British status uncertain.

C. apicalis is also distinguishable from the other two species in that it has no dark-coloured margin to the pronotum.

* see p. 13

† Zwick (1967) has split the genus *Chloroperla* on the basis of differences in the genitalia. *C. torrentium* becomes *Siphonoperla torrentium* (Pictet) and *C. apicalis* becomes *Xanthoperla apicalis* (Newman).

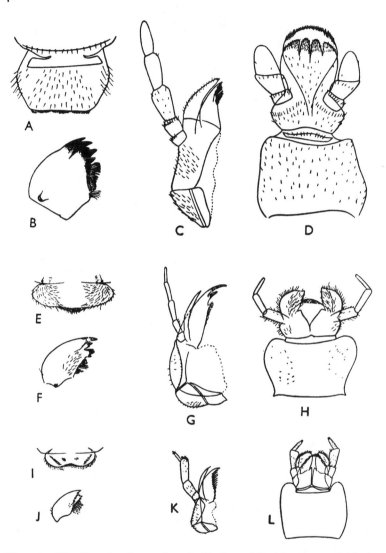

Fig. 20.   Mouthparts of nymphs: A, B, C, and D, *Nemoura avicularis*,
× 60; E, F, G, and H, *Diura bicaudata*, × 15; I, J, K, and
L, *Chloroperla torrentium*, × 30.   A, E, and I, labra; B,
F, and J, mandibles; C, G, and K, maxillae; D, H, and L,
labia with attached hypopharynx.

THE NYMPHS

These keys are only intended to apply to specimens 5 mm or more in length. Keys to smaller specimens of most species are given by Hynes (1941), but in several groups the very young stages are as yet imperfectly known.

KEY TO FAMILIES

1   Glossae as long as paraglossae (Fig. 20D). Labrum less than twice as wide as long (Fig. 20A). Mandibles short and stout (Fig. 20B). 10th sternum reduced to a narrow strip (Figs. 21E and F)—   **2**

—   Glossae reduced (Figs. 20H and L). Labrum more than twice as wide as long (Figs. 20E and I). Mandibles elongate (Figs. 20F and J). 10th sternum well developed (Fig. 41C). Tarsi with two basal segments sub-equal and much shorter than the third segment (Fig. 21A)—   **5**

2   Each segment of the tarsus longer than the preceding segment (Fig. 21B)—   TAENIOPTERYGIDAE, p. 51

—   Second segment of the tarsus shorter than the first (Figs. 21C and D)—   **3**

3   Stout nymphs with wing pads set obliquely to the body (Figs. 26A, 28A, 30A and 32A). Hind leg when stretched back alongside the abdomen greatly over-reaching its tip—   NEMOURIDAE, p. 55

—   Cylindrical elongate nymphs (Figs. 36 and 38A). Hind leg when stretched back alongside the abdomen not quite reaching its tip—   **4**

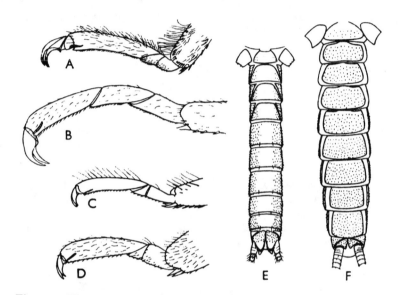

Fig. 21. Hind tarsi of nymphs: A, *Isoperla grammatica*, × 30;
B, *Taeniopteryx nebulosa*, × 30; C, *Leuctra fusca*, × 60;
D, *Capnia bifrons*, × 60.
Abdomens of nymphs in ventral view, × 12: E, *Leuctra hippopus*; F, *Capnia bifrons* (female).

4 Abdominal segments 1-4 only divided into tergum and sternum, segments 5-9 fused into complete rings. Paraprocts longer than wide (Fig. 21E)—   LEUCTRIDAE, p. 67

— Abdominal segments 1-9 divided into tergum and sternum. Paraprocts wider than long (Fig. 21F)—   CAPNIIDAE, p. 71

5(1) Pleural gills present on the thorax (Figs. 45A and 46A)—
                                                    PERLIDAE, p. 77

— No gills present on thorax—   6

6 Last segment of maxillary palp normal, more than one-quarter as wide as the preceding segment (Figs. 2 and 20G)—
                                                    PERLODIDAE, p. 73

— Last segment of maxillary palp reduced, only about one-quarter as wide as the preceding segment (Figs. 20K and 47)—
                                                    CHLOROPERLIDAE, p. 81

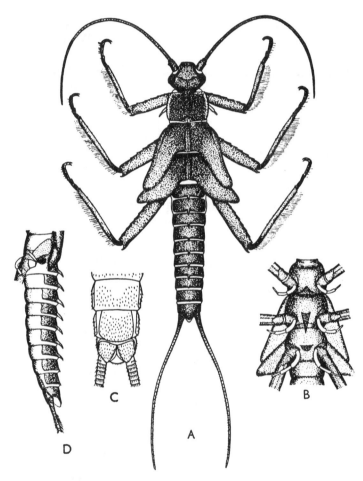

Fig. 22.   Nymph of *Taeniopteryx nebulosa*: A, full-grown female,
× 6; B, ventral view of thorax, × 6; C, ventral view of tip
of female abdomen, × 12; D, lateral view of abdomen, × 6.

Family TAENIOPTERYGIDAE

(See Hynes 1957)

1 Each coxa bearing a 3-segmented retractile filamentous gill on the inner side (Fig. 22B). Abdominal terga 1-7 each with a horn-like process posteriorly (Fig. 22D). 9th sternum not drawn out posteriorly (Fig. 22C)—   **Taeniopteryx nebulosa**
8-12 mm (p. 23) (see Aubert 1950)

— Coxae without gills. Abdominal terga without processes. 9th sternum drawn out into a large posteriorly projecting lobe (Figs. 23 and 25B and c)—   **2**

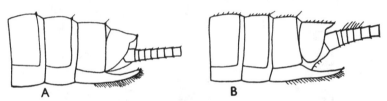

Fig. 23.   Tips of abdomens of female nymphs in lateral view, × 16: A, *Rhabdiopteryx acuminata*; B, *Brachyptera putata*. Only those hairs and bristles which can be seen in profile are shown.

2 Cerci without long hairs on the upper side. Clothing hairs of abdominal terga lying flat, so that they are not seen in profile (Fig. 23A). Colour pattern as in Fig. 24A and D—
Rhabdiopteryx acuminata
7-10 mm (p. 23)

— Cerci with long hairs on the upper sides of the basal few segments. Clothing hairs on abdominal terga bristle-like and upstanding, so that they are clearly seen in profile (Fig. 23B)—
genus BRACHYPTERA   **3**

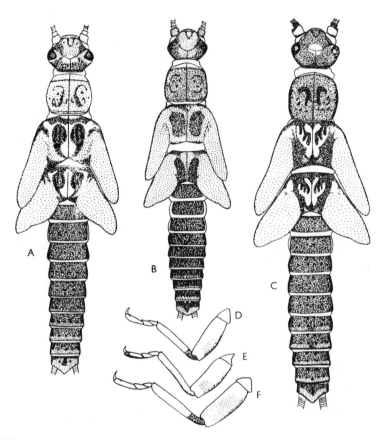

Fig. 24. Fully grown female nymphs without legs, × 8·5, and fore-
legs, × 10: A, D, *Rhabdiopteryx acuminata*; B, E,
Brachyptera risi; C, F, *B. putata*.

3 Tarsi pale; tibiae with a clearly defined dark area at the base
(Fig. 24F). Abdominal terga each with a transverse line of
darker spots, the 10th being without pale spots. Meso- and
meta-notal pattern complex (Fig. 24C)— **Brachyptera putata**
7-11 mm (p. 25)

— Tarsi darkened; tibiae without a clearly defined dark area at
the base (Fig. 24E). Abdominal terga without a transverse
line of darker spots, the 10th with two small but distinct pale
spots on its anterior border. Meso- and meta-notal pattern
simple (Figs. 24B and 25A)— **Brachyptera risi**
7-10 mm (p. 25)

Well-grown male nymphs of *Rhabdiopteryx* and the two species
of *Brachyptera* are easily distinguishable on the shape of the
paraprocts. In *R. acuminata* these end in blunt points, in *B.
putata* their tips are drawn out into horns which are turned
sideways at an angle of about 45° to the axis of the body, while
in *B. risi* these horns are so twisted that their tips point quite
laterally or even slightly forwards (Fig. 25B).

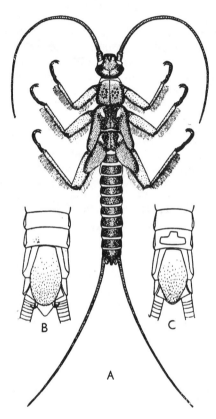

Fig. 25. Nymph of *Brachyptera risi*: A, full-grown male, × 6. B, C, ventral view of tips of abdomens, × 12: B, male; C, female.

Family NEMOURIDAE
(See Kühtreiber 1934, Hynes 1941, 1963, Brinck 1949,
Wojtas 1963, Raušer 1956, 1963)

1 Prosternum with three sausage-shaped gills on each side (Fig. 26D)— genus PROTONEMURA 2
(Raušer 1956)

— Prosternum with two bunches of 5-8 filamentous gills on each side (Fig. 29D). Nymphs usually heavily coated with flocculent detritus— genus AMPHINEMURA 3

— No prosternal gills— 4

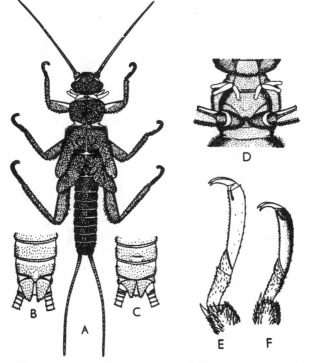

Fig. 26. Nymphs of *Protonemura*: A, full-grown *P. meyeri*, × 6;
B, ventral view of abdomen of male *P. meyeri*, × 12;
C, ventral view of abdomen of female, × 12; D, prothorax
in ventral view, × 12; E, hind tarsus of *P. praecox*, × 30;
F, hind tarsus of *P. meyeri*, × 30.

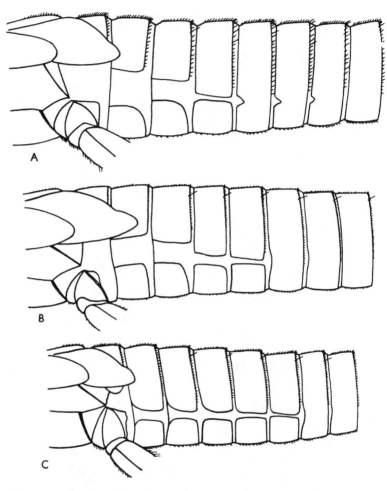

Fig. 27.   Abdomens of *Protonemura* nymphs in side view, × 30:
A, *P. meyeri*; B, *P. praecox*; C, *P. montana*.
Only those bristles are shown which form the posterior
fringes to the terga or are visible in profile.

2   General colour dark green-grey. Abdominal segments 1-4 divided into tergum and sternum—segments 5-8 usually show partial division, but this is never complete (Fig. 27A). Abdominal terga with a simple fringe of bristles on the posterior margin, one or two of which may be longer than the others, but never more than twice as long (Fig. 27A). Paraprocts of fully grown male slightly emarginate at the tip (Fig. 26B). Tarsi darkened on the upper side (Fig. 26F), antennae darkened distally, distal cercal segments darkened at the tips and bases so that the cercus is ringed (Fig. 26A)—

Protonemura meyeri
8-10 mm (p. 28)

—   General colour rufous brown. Abdominal segments 1-5 divided into tergum and sternum—segments 6-8 without partial division (Fig. 27B). Abdominal terga 2 or 3 to 7, 8 or 9 bearing a long dorsolateral bristle on each side of the posterior margin which is two or three times as long as the other bristles in the posterior fringe (Fig. 27B). Paraprocts of fully grown male simple and rounded at the tip. Tarsi (Fig. 26E), antennae and cerci uniformly yellow. Nymphs large in winter and spring—   **Protonemura praecox**
8-10 mm (p. 27)

—   General colour yellowish-brown. Abdominal segments 1-6 divided into tergum and sternum—segments 7 and 8 without partial division (Fig. 27C). Abdominal terga 3 to 7 or 8 bearing a long dorsolateral bristle on each side of the posterior margin which is two or three times as long as the other bristles in the posterior fringe (Fig. 27C). Paraprocts of fully grown male pointed at the tip. Tarsi, antennae and cerci uniformly yellowish-brown. Nymphs large in summer and autumn—

Protonemura montana
8-10 mm (p. 28) (Kimmins 1943)

3(1) General colour usually chocolate brown, paler in smaller
specimens. Bristles on the femora sharply differentiated into
large and small, the larger bristles being arranged in a group
about two-thirds down the length of each femur (Fig. 29F)
(best seen in hind leg). Segments of the cerci enlarged towards
the tip; bristles in whorls on middle segments of the cerci
(segments 8 to 14) only about half as long as the segments
except for an occasional longer bristle (Fig. 28A)—

Amphinemura sulcicollis

4-6 mm (p. 29)

— General colour orange yellow, yellow in smaller specimens.
Bristles on the femora not sharply differentiated into two types
and more or less evenly distributed (Fig. 29E). Segments of
the cerci cylindrical, only those at the tip showing any sign
of bulbous ends; bristles in whorls on middle segments of the
cerci (segments 8 to 14) about two-thirds the lengths of the
cerci (Fig. 28B)—   **Amphinemura standfussi**

4·5-6 mm (p. 29)

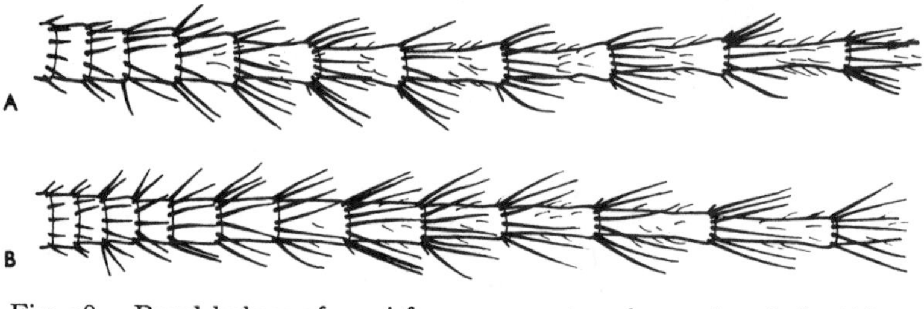

Fig. 28.   Basal halves of cerci from segment 3 of nymphs of *Amphinemura*, × 60: A, *A. sulcicollis*; B, *A. standfussi*.

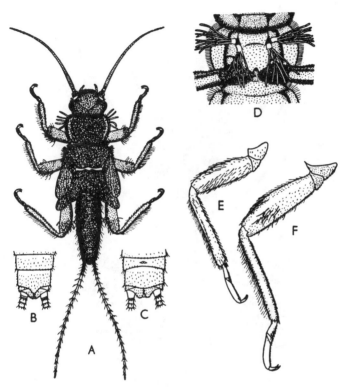

Fig. 29. Nymphs of *Amphinemura*: A, full-grown *A. sulcicollis*, ×9;
B, ventral view of abdomen of male, × 18; C, ventral
view of abdomen of female, × 18; D, prothorax in ventral
view, × 18; E, hind leg of *A. standfussi*, × 18; F, hind
leg of *A. sulcicollis*, × 18.

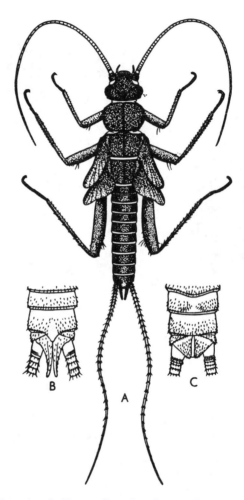

Fig. 30. Nymph of *Nemurella picteti*: A, full-grown male, × 9;
B, ventral view of abdomen of male, × 18; C, ventral
view of abdomen of female, × 18.

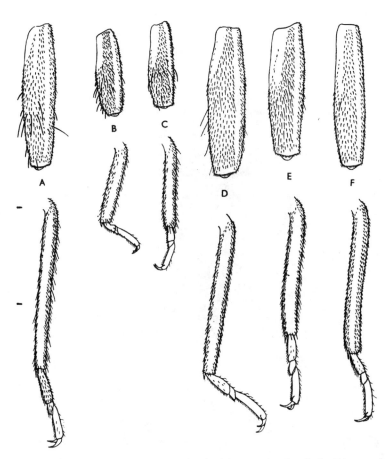

Fig. 31. Right hind femora in dorsal view and left hind tibiae and tarsae of nymphs, × 21·5: A, *Nemurella picteti*; B, *Nemoura cambrica*; C, *N. erratica*; D, *N. avicularis*; E, *N. cinerea*; F, *N. dubitans*.

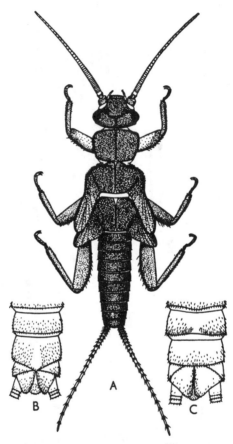

Fig. 32.   Nymphs of *Nemoura*: A, full-grown *N. erratica*, × 9; B,
ventral view of abdomen of male, × 18; C, ventral view
of abdomen of female, × 18.

4(1) Segments 1 and 3 of hind tarsus of about equal length (Fig. 31A). Hind and other femora with a well-defined transverse row of long stout bristles on the upper side, the longest about as long as the width of the femur (Figs. 30 and 31A). (Cercal bristles very stout and long (Fig. 33A); dorso-lateral tergal bristles of middle abdominal segments longer than the widths of their terga (Fig. 35A))— **Nemurella picteti**

7-9 mm (p. 31)

— Segment 1 of hind tarsus about one-third to one-half the length of segment 3 (Fig. 31B-F). Hind femora without a transverse row of bristles (Fig. 31B, D and E), or if such a row is present the bristles are few and short (Fig. 31F) or the row is ill-defined (Fig. 31C)— genus NEMOURA **5**

5 Stout bodied species with relatively short legs (Fig. 32). Tibia of hind leg 6-8 times as long as wide; femora with many long stout curved bristles (Fig. 31B and C). Cercal segments usually without intermediate hairs, although a few may be present on very distal segments (Fig. 33B and C)— **6**

— More slender species with longer legs. Tibia of hind leg 9-11 times as long as wide; femora with fine straight bristles, some of which may be long (Fig. 31D, E and F). Cercal segments with intermediate hairs (Fig. 33D, E and F)— **7**

6 Tentorial callosities crescentic (Fig. 34A). Bristles on hind femora evenly scattered, of intergrading lengths, the longest up to nearly half the width of the femur (Fig. 31B). Dorso-lateral tergal bristles of middle abdominal segments up to half as long as the tergum is wide (Fig. 35B). Cercal bristles about half as long as the segment which is 3 times as long as wide (Fig. 33B)— **Nemoura cambrica**

5-6 mm (p. 32)

— Tentorial callosities more or less triangular (Fig. 34B). Bristles on femora of two types, short and evenly scattered, and long, up to more than half the width of the femur, which tend to form an irregular double transverse row across the femur (Fig. 31C). Dorsolateral tergal bristles about three-quarters as long as the tergum is wide (Fig. 35C). Cercal bristles about three-quarters as long as the segment which is 3 times as long as wide (Fig. 33C). (Note—in specimens from high altitudes all bristles may be shorter than indicated here)— **Nemoura erratica**

6·5-9 mm (p. 32)

7(5) Second antennal segment much darker coloured than first (Fig. 34C). Dorsolateral bristles (of which there may be 2 or 3 series) about three-quarters as long as the width of the terga (Fig. 35D). Colour malt olive-grey. (Cercal bristles fine and mostly longer than the segments (Fig. 33D))—
Nemoura avicularis
7-8 mm (p. 32)

— Second antennal segment not darker coloured than first (Fig. 34D). Dorsolateral bristles less than three-quarters as long as the width of the terga (Fig. 35E and F). Colour pale brown—
8

8 Ocelli clearly visible as dark spots (Fig. 34D). Hind femoral bristles of varying intergrading lengths and evenly scattered; tibial bristles up to half as long as the width of the tibia (Fig. 31E). Dorsolateral tergal bristles of middle abdominal segments about half as long as the width of the terga (Fig. 35E). Cercal bristles clearly darker than cercus, fairly stout and up to three-quarters as long as the segment which is 3 times as long as wide; intermediate hairs short (Fig. 33E)—
Nemoura cinerea
7-10 mm (p. 31)

— Ocelli visible but not darkened. Hind femoral bristles stout and short, except for a few which form a single row across the top of the femur; tibial bristles short, curved, and less than one-third as long as the width of the tibia (Fig. 31F). Dorsolateral tergal bristles very short, one-quarter as long as width of terga, and sometimes no longer than other bristles of the posterior tergal fringe (Fig. 35F). Cercal bristles not darker than cercus, fine, and as long as the segment which is 3 times as long as wide; intermediate hairs long, particularly on distal segments (Fig. 33F)— **Nemoura dubitans**
6-9 mm (p. 32)

Fig. 33. Basal halves of cerci of nymphs, × 49: A, *Nemurella picteti*; B, *Nemoura cambrica*; C, *N. erratica*; D, *N. avicularis*; E, *N. cinerea*; F, *N. dubitans*.

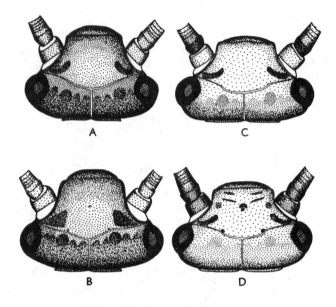

Fig. 34. Nymphs of *Nemoura*, heads, × 35: A, *N. cambrica*;
B, *N. erratica*; C, *N. avicularis*; D, *N. cinerea*.

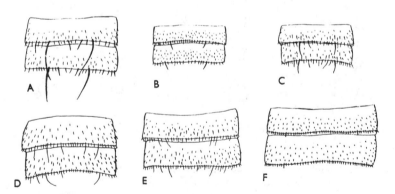

Fig. 35. Fourth and fifth abdominal terga of nymphs, × 24·5: A,
Nemurella picteti; B, *Nemoura cambrica*; C, *N. erratica*; D,
N. avicularis; E, *N. cinerea*; F, *N. dubitans*.

Family LEUCTRIDAE
Genus LEUCTRA
(See Kühtreiber 1934, Hynes 1941, Illies 1955)

1 Head broad and flattened (Fig. 36A). Antennae stout and curved, with chitinous processes and long bristles on the basal segments (Fig. 37 I). Pronotum with a thick fringe of long hairs: general colour grey-brown—   **Leuctra geniculata**
8-11 mm (p. 33)

— Head rounded (Fig. 36B and C). Antennae slender and filiform (Fig. 37J)—   **2**

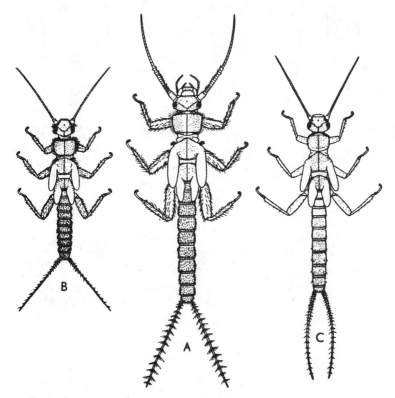

Fig. 36. Full-grown nymphs of *Leuctra*, × 6; A, *L. geniculata*; B, *L. nigra*; C, *L. hippopus*.

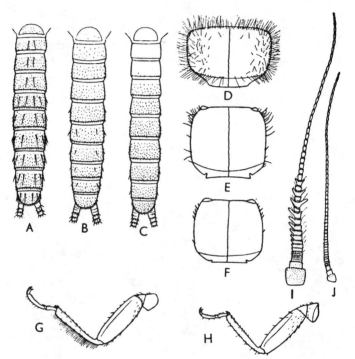

Fig. 37.   Nymphs of *Leuctra*, A-C, abdomens in dorsal view, × 12:
A, *L. inermis*; B, *L. hippopus*; C, *L. moselyi*;
D-F, pronota, × 24; D, *L. nigra*; E, *L. hippopus*;
F, *L. moselyi*;
G, H, hind legs, × 12; G, *L. fusca*; H, *L. hippopus*;
I, J, antennae, × 12; I, *L. geniculata*; J, *L. hippopus*.

2 Pronotum with a thick fringe of long fine hairs (Fig. 37D), and similar hairs over most of the body (Fig. 36B). General colour orange yellow—   **Leuctra nigra**
6-7 mm (p. 35)

— Pronotum with a few bristles laterally (Fig. 37E and F), and a few fine bristles on the rest of the body (Fig. 36C). General colour greyish yellow—   **3**

3 Tibiae of all legs with a fringe of long fine hairs on the posterior margin (Fig. 37G)—   **Leuctra fusca**
6.5-9 mm (p. 35)

— Tibiae of all legs with only a few scattered fine hairs on the posterior margin, not forming a fringe (Fig. 37H)—   **4**

4 Abdominal terga 2-10 each with at least one pair of long dorsolateral bristles (Fig. 37A)—   **Leuctra inermis**
6-8 mm (p. 33)

—·— Abdominal terga 1-4 without long dorsolateral bristles (Figs. 37B and C) although they may be present on the more posterior terga (Fig. 37B)—   **5**

5 Specimens longer than 5 mm collected between October and May inclusive. More than 12 bristles on each side of the pronotum (Fig. 37E). Clothing hairs present on abdominal tergum 2 and usually also on tergum 1 (Fig. 37B). (The pronotal bristles are often broken and the clothing hairs difficult to see)—   **Leuctra hippopus**
6-8.5 mm (p. 33)

— Specimens longer than 5 mm collected between May and September inclusive. Less than 12 bristles on each side of the pronotum (Fig. 37F). No clothing hairs on abdominal terga 1 and 2 (Fig. 37C)—   **Leuctra moselyi**
6-8.5 mm (p. 35)

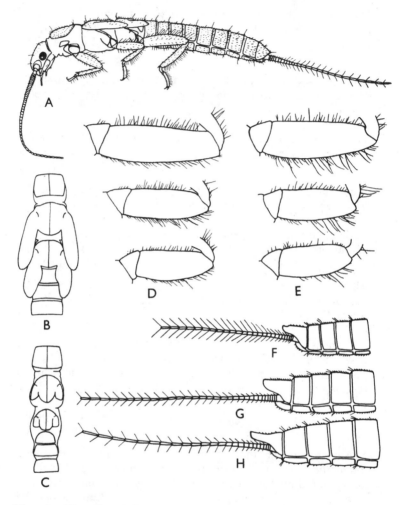

Fig. 38.   Nymphs of *Capnia*: A, full-grown female *C. bifrons*, × 10;
B, thorax of full-grown male *C. atra*, × 10; C, thorax
of full-grown male *C. vidua*, × 10; D, femora of *C. vidua*;
E, femora of *C. bifrons*;
F-H, ends of abdomens and basal two-thirds of cerci of
full-grown male nymphs in lateral view, × 15: F, *C. atra*;
G, *C. vidua*; H, *C. bifrons*.
In D, E, F, G, and H, only those hairs and bristles which
can be seen in profile are shown.

Family CAPNIIDAE
Genus CAPNIA
(See Brinck 1949, Hynes 1955)

1 Cerci with a dorsal and ventral fringe of long hairs which in the middle of each cercus are longer than the segments which bear them (Fig. 38F). Wing-pads of well-grown male nymph fully developed (Fig. 38B). 10th tergum of fully grown male nymph pointed in dorsal and lateral view and emarginate dorsally (Fig. 38F). 8th sternum of ripe female nymph with a heavily chitinised vagina. Femora and abdominal terga as in *C. vidua,* Fig. 38D—   **Capnia atra**
6-8 mm (p. 37)

— Hairs of dorsal and ventral cercal fringes shorter, being in the middle of each cercus shorter than or as long as the segments which bear them (Figs. 38G and H). Wing-pads of well-grown male nymph rudimentary (Fig. 38C). 10th tergum of fully grown male nymph not pointed or emarginate. Female nymph without a heavily chitinised vulva—   **2**

2 Each abdominal tergum with a pair of short upstanding hairs about half-way along the mid-dorsal line (Figs. 38A and H). 10th tergum of fully grown male nymph truncated in lateral view (Fig. 38H). Bristles along posterior borders of femora longer than those along anterior borders (Fig. 38E)—
Capnia bifrons
6-9 mm (p. 37)

— No short upstanding hairs half-way along abdominal terga (Fig. 38G). 10th tergum of fully grown male nymph broadly rounded posteriorly (Fig. 38G). Bristles along posterior borders of femora about the same length as those along anterior borders (Fig. 38D)—   **Capnia vidua**
6-8 mm (p. 37)

Fig. 39.   Lateral views of bases of abdomens of nymphs,   × 5:
A, *Perlodes microcephala*; B, *Diura bicaudata*.

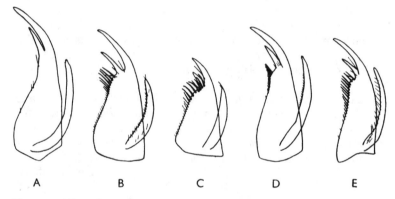

Fig. 40.   Tips of maxillae (laciniae and galeae) of nymphs : A, *Perlodes
microcephala*,   × 21;   B,   *Isoperla grammatica*,   × 40;
C, *I. obscura*, × 40; D, *Diura bicaudata*, × 21; E, *Isogenus
nubecula*, × 21 (from a Swedish specimen).

Family PERLODIDAE
(See Hynes 1941, Brinck 1949, Illies 1956)

1 Abdominal segments 1-4 divided into tergum and sternum (Fig. 39A). Lacinia narrowing evenly towards the tip, with only a few fine scattered bristles on the inner margin (Fig. 40A)—   **Perlodes microcephala**
18-28 mm (p. 41) (See Fig. 42)

— Abdominal segments 1 and 2 only divided into tergum and sternum, segments 3 and 4 fused into a complete ring (Fig. 39B). Lacinia more or less abruptly narrowed just below the tip and fringed with stout bristles on the inner margin (Fig. 40B-E)—   **2**

2 Paraprocts blunt (Figs. 43B and c). Body sparsely covered with scattered bristles, and sometimes also with clothing hairs which lie flat against the body—   **4**

— Paraprocts long and pointed (Fig. 41C). Body thickly covered with black clothing hairs, which are particularly obvious on pale areas; stouter upstanding bristles always absent—
genus ISOPERLA **3**

3 Galea with a row of hairs on the inner margin; stout bristles on inner margin of lacinia arranged in a single row (Fig. 40B)—
Isoperla grammatica
11-16 mm (p. 42) (See Fig. 41)

— Galea without hairs on the inner margin; stout bristles on inner margin of lacinia arranged in a double row (Fig. 40C)—
Isoperla obscura*†
9-13 mm (p. 42)

A further, unreliable, difference between these two species is to be found in the pattern on the head. In *I. obscura* there is a pale spot in the centre of the posterior arm of the epicranial suture, whereas in *I. grammatica* there are two spots, one on each side of the suture, which runs across a dark area (Fig. 41A). These two spots may, however, enlarge and fuse and even come to occupy the whole vertex (Fig. 41B), in which event the suture crosses a pale area. Also in small specimens the pattern is not developed.

* see p. 13

† see amendment on p. 91.

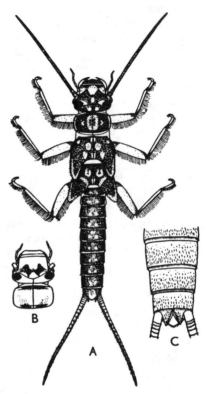

Fig. 41. Full-grown nymph of *Isoperla grammatica* : A, dorsal view,
× 5; B, head and pronotum of very pale specimen, × 5;
C, central view of abdomen tip, × 10.

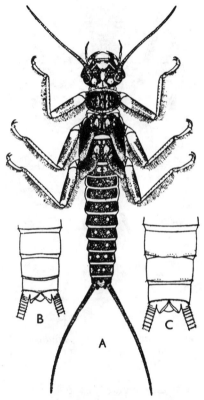

Fig. 42. Full-grown nymph of *Perlodes microcephala*: A, female, × 2·5; B, ventral view of male abdomen, × 5; C, ventral view of female abdomen, × 5.

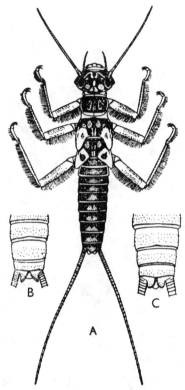

Fig. 43. Full-grown nymph of *Diura bicaudata*: A, female, × 2·5; B, ventral view of male abdomen, × 5; C, central view of female abdomen, × 5.

4(2) Lacinia strongly emarginate just below the sub-apical tooth, with a resulting gap between this tooth and the inner fringe of bristles (Fig. 40D). No clothing hairs—   **Diura bicaudata**
8·5-17 mm (p. 41) (See Fig. 43)

— Lacinia only slightly notched below the sub-apical tooth with a very slight gap between the tooth and the fringe of bristles (Fig. 40E). Clothing hairs present between the bristles on the dorsal surface—   **Isogenus nubecula**
14-21 mm (p. 39)

Perlodes microcephala and *Diura bicaudata* are the two common large perlodids found in this country, and are readily distinguished by the differences in colour pattern illustrated in Figs. 42A and 43A. The pattern of *Isogenus nubecula* resembles that of *Diura bicaudata*.

Family PERLIDAE

(See Schoenemund 1925, Hynes 1941, Aubert 1949, Illies 1955)

1 Pronotum more than twice as wide as long (Fig. 45A). Sub-mentum with anterior corners separated off by sutures (Fig. 44A). General colour red-brown to dark brown with a pattern of yellow or pale grey, last abdominal tergum uniformly dark (Fig. 45A)— **Dinocras cephalotes**

14-31 mm (p. 43)

— Pronotum less than twice as wide as long (Fig. 46A). Anterior corners of sub-mentum not separated off by sutures (Fig. 40B). General colour black with a bold yellow pattern, last abdominal tergum yellow (Fig. 46A)— **Perla bipunctata**

16-33 mm (p. 43)

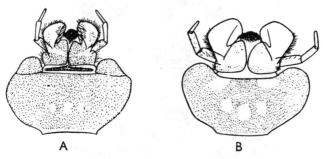

A   B

Fig. 44. Labia of full-grown nymphs, with attached hypopharynx, × 7·5: A, *Dinocras cephalotes*; B, *Perla bipunctata*.

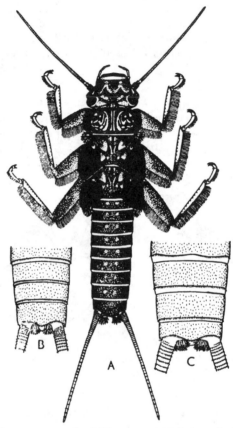

Fig. 45. Full-grown nymph of *Dinocras cephalotes*: A, female, ×2·5; B, ventral view of male abdomen, × 5; C, ventral view of female abdomen, × 5.

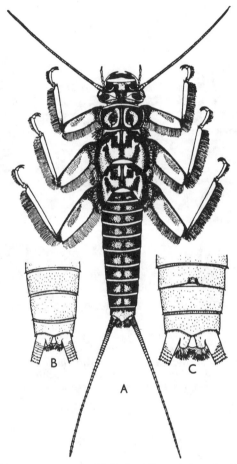

Fig. 46   Full-grown nymph of *Perla bipunctata*: A, female, × 2·5;
B, ventral view of male abdomen, × 5; C, ventral view of
female abdomen, × 5.

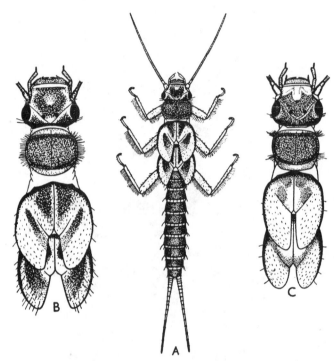

Fig. 47. Full-grown nymphs of *Chloroperla*: A, *C. torrentium*,
× 7·5; B, head and thorax of *C. torrentium*, × 15; C,
head and thorax of *C. tripunctata*, × 15.

Family CHLOROPERLIDAE
Genus CHLOROPERLA†
(See Hynes 1941, Aubert 1951, Illies 1955)

Chloroperla apicalis*† (for habitat see p. 45) is not included in the following key because no specimens have been available. Aubert (1953), however, describing specimens from Italy, states that the species is easily distinguishable from all other known nymphs of the genus. It is smaller (5-7 mm), much paler coloured, with longer wing-pads, and the abdomen is clearly banded, each tergum being darkened anteriorly and pale posteriorly (cf. Fig. 47A). Elsewhere darkened areas occur only on the mesonotum, pronotum, and back of the head, with occasionally a little darkening in front of the M-line. The area between the ocelli, which is darkened in the two certainly British species, is pale yellow. The pronotal bristles are arranged as in *C. tripunctata* (Fig. 47C).

1   M-line conspicuous as a pale line or as the boundary between a pale and dark area (Fig. 47C). Posterior arm of epicranial suture about as long as the distance from the fork to each lateral ocellus (Fig. 47C). Pronotum with long fine bristles arranged in anterior and posterior groups on each side (Fig. 47C)—   **Chloroperla tripunctata**
8-10 mm (p. 45)

—   M-line inconspicuous (Figs. 47A and B). Posterior arm of epicranial suture very short or absent (Fig. 47B). Pronotum with an almost continuous fringe of long fine bristles round each lateral margin (Fig. 47B)— **Chloroperla torrentium**†
7-9 mm (p. 45)

* see p. 13
† see footnote on p. 45

DISTRIBUTION MAPS

The vice-counties from which each species has been certainly recorded are blackened. Species about whose distribution little is known or which have very restricted distributions are discussed on pp. 16-17; they are *Rhabdiopteryx acuminata, Isogenus nubecula, Isoperla obscura* and *Chloroperla apicalis*.

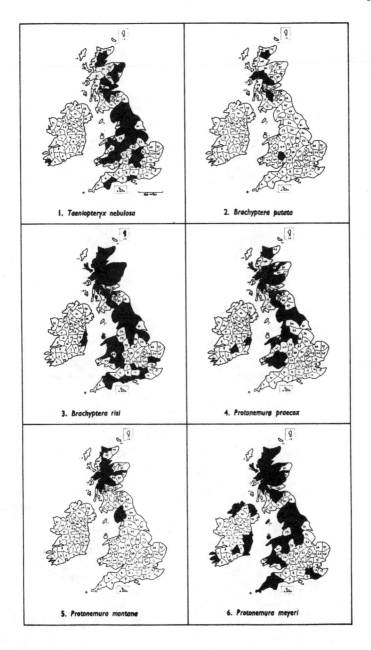

1. Taeniopteryx nebulosa

2. Brachyptera putata

3. Brachyptera risi

4. Protonemura praecox

5. Protonemura montana

6. Protonemura meyeri

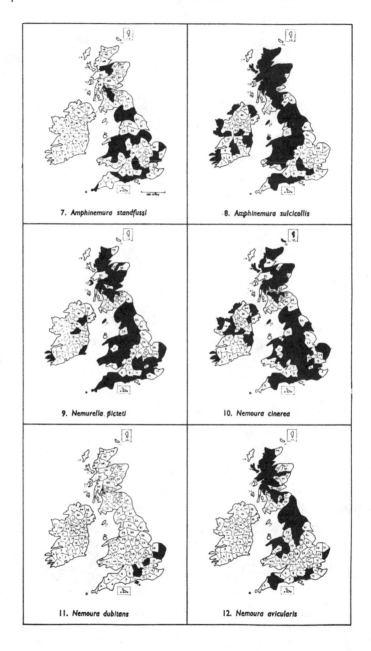

7. Amphinemura standfussi

8. Amphinemura sulcicollis

9. Nemurella pícteti

10. Nemoura cinerea

11. Nemoura dubitans

12. Nemoura avicularis

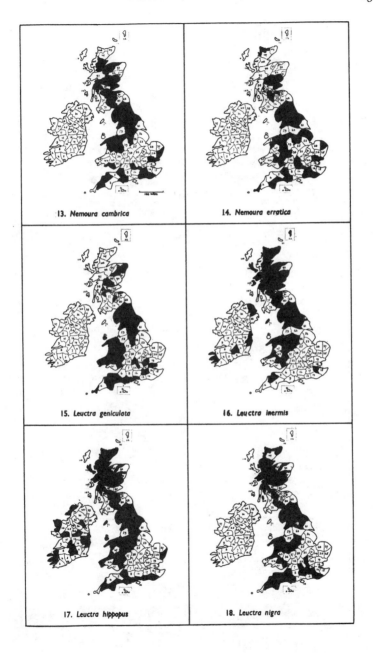

13. *Nemoura cambrica*

14. *Nemoura erratica*

15. *Leuctra geniculata*

16. *Leuctra inermis*

17. *Leuctra hippopus*

18. *Leuctra nigra*

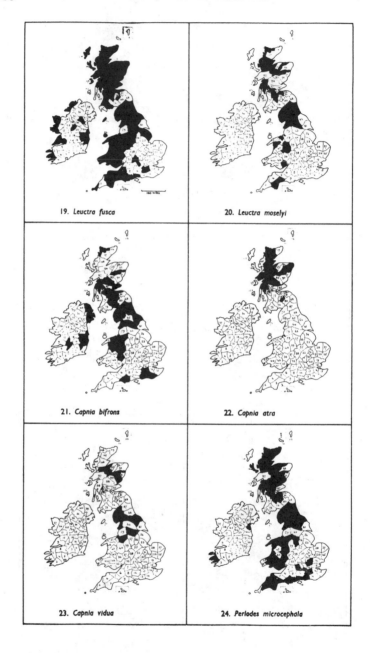

19. Leuctra fusca

20. Leuctra moselyi

21. Capnia bifrons

22. Capnia atra

23. Capnia vidua

24. Perlodes microcephala

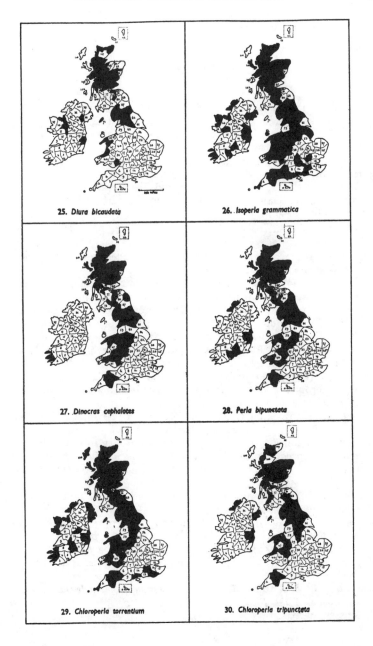

25. *Diura bicaudata*

26. *Isoperla grammatica*

27. *Dinocras cephalotes*

28. *Perla bipunctata*

29. *Chloroperla torrentium*

30. *Chloroperla tripunctata*

REFERENCES

The list of references includes those cited in the introduction and general works which contain modern information on the order. Almost all the papers on the group published before 1950 are listed by Claassen (1940) and Hanson & Aubert (1952). The references in the keys are to modern papers giving good comparative descriptions and figures of the adults or nymphs of the family, genus or species after which the reference is cited. It should be emphasized here that identification, especially of nymphs, cannot be certain without reference to such a description.

Aubert, J. (1945). Le microptérisme chez les Plécoptères (Perlariés). *Revue suisse Zool.* **52**, 395-99.

Aubert, J. (1949). Plécoptères helvétiques. Notes morphologiques et systématiques. *Mitt. schweiz. ent. Ges.* **22**, 217-35.

Aubert, J. (1950). Note sur les Plécoptères européens du genre *Taeniopteryx* Pictet (*Nephelopteryx* Klapalek) et sur *Capnia vidua* Klapalek. *Mitt. schweiz. ent. Ges.* **23**, 303-16.

Aubert, J. (1951). Plécoptères helvétiques: description de larves nouvelles. *Mitt. schweiz. ent. Ges.* **24**, 279-98.

Aubert, J. (1953). Contribution à l'étude des Plécoptères et des Ephémeroptères de la Calabre (Italie méridionale). *Ann. Ist. Mus. zool. Napoli,* **5** (2), 1-35.

Aubert, J. (1956). Contribution à l'étude des Plécoptères d'Afrique du Nord. *Mitt. schweiz. ent. Ges.* **29**, 419-36.

Balfour-Browne, F. (1931). A plea for uniformity in the method of recording insect captures. *Entomologist's mon. Mag.* **67**, 183-93.

Brinck, P. (1949). Studies on Swedish stoneflies. *Opusc. ent.* Suppl. 11, 250 pp.

Brinck, P. (1956). Reproductive system and mating in Plecoptera. *Opusc. ent.* **21**, 57-127.

Claassen, P. W. (1940). A catalogue of the Plecoptera of the world. *Mem. Cornell Univ. agric. Exp. Stn,* No. 232, 235 pp.

Comstock, J. H. (1918). *The wings of insects.* Ithaca, N.Y.

Despax, R. (1951). Plécoptères. *Faune de France,* **55**, 280 pp. Paris.

Hanson, J. F. & Aubert, J. (1952). *First supplement to the Claassen Catalogue of the Plecoptera of the world.* 23 pp. University of Massachussetts.

Hynes, H. B. N. (1941). The taxonomy and ecology of the nymphs of British Plecoptera with notes on the adults and eggs. *Trans. R. ent. Soc. Lond.* 91, 459-557.

Hynes, H. B. N. (1942). A study of the feeding of adult stoneflies (Plecoptera). *Proc. R. ent. Soc. Lond.* (A) 17, 81-2.

Hynes, H. B. N. (1955). The nymphs of the British species of *Capnia* (Plecoptera). *Proc. R. ent. Soc. Lond.* (A) 30, 91-6.

Hynes, H. B. N. (1957). The British Taeniopterygidae (Plecoptera). *Trans. R. ent. Soc. Lond.* 109, 233-43.

Hynes, H. B. N. (1962). The hatching and growth of the nymphs of several species of Plecoptera. *Proc. XI int. Congr. Ent. Vienna* 3, 271-3.

Hynes, H. B. N. (1963a). *Isogenus nubecula* in Britain (Plecoptera : Perlodidae). *Proc. R. ent. Soc. Lond.* (A) 38, 12-14.

Hynes, H. B. N. (1963b). The gill-less Nemourid nymphs of Britain (Plecoptera). *Proc. R. ent. Soc. Lond.* (A) 38, 70-6.

Illies, J. (1952). Die europäischen Arten der Plecopterengattung *Isoperla* Banks (= *Chloroperla* Pictet). *Beitr. Ent.* 2, 369-424.

Illies, J. (1953). Beitrag zur Verbreitungsgeschichte der europäischen Plecopteren. *Arch. Hydrobiol.* 48, 35-74.

Illies, J. (1955a). Steinfliegen oder Plecopteren. *Tierwelt Dtl.* 43, 150 pp. Jena.

Illies, J. (1955b). Die Bedeutung der Plecopteren für die Verbreitungsgeschichte der Süsswasserorganismen. *Verh. int. Verein. theor. angew. Limnol.* 12, 643-53.

Illies, J. (1960). Verbreitungsgeschichte der Plecopteren auf der Südhemisphare. *Proc. XI int. Congr. Ent. Vienna* 1, 476-80.

Illies, J. (1965). Phylogeny and zoogeography of the Plecoptera. *A. Rev. Ent.* 10, 117-40.

Kaslauskas, R. (1962). Kai kurie duomenys apie Lietuvos tsr ankstyves (Plecoptera). (Lithuanian : Russian summary). *Lietuvos TSR aukstuju mokyklu Mokslo Darbai:* Biologija 2, 168-74.

Khoo, S. G. (1964). Studies on the biology of *Capnia bifrons* (Newman) and notes on the diapause of the nymphs of this species. *Gewäss. Abwäss.* 34/35, 23-30.

Kimmins, D. E. (1936). Synonymic notes on the genera *Chloroperla, Isopteryx* and *Isoperla* (Plecoptera), with a list of the British species of *Chloroperla. J. Soc. Br. Ent.* 1, 121-4.

Kimmins, D. E. (1940). A synopsis of the British Nemouridae (Plecoptera). *Trans. Soc. Br. Ent.* 7, 65-83.

Kimmins, D. E. (1941). A new species of Nemouridae (Plecoptera). *J. Soc. Br. Ent.* 2, 89-93.

Kimmins, D. E. (1943). *Rhabdiopteryx anglica,* a new British species of Plecoptera. *Proc. R. ent. Soc. Lond.* (B) 12, 42-4.

Kimmins, D. E. (1943). The nymph of *Protonemura montana* (Plecoptera). *J. Soc. Br. Ent.* 2, 159.

Kimmins, D. E. (1944). The Plecoptera in the Dale collection of British insects. *Entomologist's mon. Mag.* **80,** 273-8.

Kimmins, D. E. (1950). Plecoptera. *Handbooks for the Identification of British Insects.* **I, 6,** 18 pp. London.

Kohtreiber, J. (1934). Die Plekopterenfauna Nordtirols. *Ber. naturw.-med. Ver. Innsbruck,* **44,** 1-219.

Morton, K. J. (1911). On *Taeniopteryx putata* Newman (Plecoptera), with notes on other species of the genus. *Entomologist,* **44,** 81-7.

Morton, K. J. (1913). An addition to the list of British Plecoptera: reinstatement of *Chloroperla venosa. Entomologist,* **46,** 73-7.

Morton, K. J. (1929). Notes on the genus *Leuctra* with descriptions of two new species, and on the genus *Capnia* including a species new to the British fauna. *Entomologist's mon. Mag.* **65,** 128-34.

Morton, K. J. (1934). What is *Phryganea bicaudata* of Linné? *J. Soc. Br. Ent.* **I,** 42-3.

Mosely, M. E. (1932). A revision of the European species of the genus *Leuctra. Ann. Mag. nat. Hist.* (10) **10,** 1-40.

Percival, E. & Whitehead, H. (1928). Observations on the ova and oviposition of certain Ephemeroptera and Plecoptera. *Proc. Leeds phil. lit. Soc.* **I,** 271-88.

Rauŝer, J. (1956). Zur Kenntniss der tchechoslowakischen *Protonemura*-Larven. *Pr. brn. Zakl. čsl. Akad. Ved,* **28,** 449-98.

Rauŝer, J. (1962). Zur Verbreitungsgeschichte einer Insektendauergruppe (Plecoptera) in Europa. *Pr. brn. Zakl. čsl. Akad. Ved,* **24,** 281-383.

Rauŝer, J. (1963). Contribution à la connaissance des larves du genre *Amphinemura* de la Tchécoslovaquie (Plecoptera). *Cas. čsl. Spol. ent.* **60,** 32-54.

Ricker, W. E. (1952). Systematic studies in Plecoptera. *Indiana Univ. Publs Sci. Ser.* No. 18, 200 pp.

Schoenemund, E. (1924). Plecoptera. *Biologie Tiere Dtl.* **32,** 34 pp.

Schoenemund, E. (1925). Die Larven der deutschen *Perla*-Arten (Plecoptera). *Ent. Mitt.* **14,** 113-21.

Sharov, A. G. (1960). The origin of the order Plecoptera. *Proc. XI int. Congr. Ent. Vienna* **I,** 296-8.

Wojtas, F. (1963). Beschreibung der bisher unbekannten Plecopterenlarve von *Nemoura dubitans* Morton 1894 (Plecoptera). *Mitt. schweiz. ent. Ges.* **35,** 284-7.

Wu, C. F. (1923). Morphology, anatomy and ethology of *Nemoura. Bull. Lloyd. Libr.,* (Ent. Ser.) **3,** 1-46.

APPENDIX

Third Edition — Amendment of key to nymphs of *Isoperla* (p. 73).

Hynes and Macphee (1970) have re-examined the maxillary characters of *Isoperla grammatica* and *I. obscura* and found that in small specimens of the former there are no hairs on the galea and that there is "a distinct lower row of three stout bristles, which steadily become hair-like until, when the nymph is 5-6 mm long, they can be described as hairs. The maxilla of *I. grammatica* therefore resembles that described for *I. obscura* until the nymph is about 5 mm long.

However, study of a number of specimens of *I. obscura* from Sweden, and of the one British specimen, revealed a consistent difference in the laciniae of the two species. In *I. obscura* there are four bristles in the lower row, and they remain as stout bristles right through nymphal development". It is concluded that "the two species can be distinguished, when larger than about 6 mm long, on the presence or absence of galeal hairs and the presence of hairs, as opposed to bristles, below and proximal to the subapical tooth. In smaller specimens of both there are no galeal hairs, and the bristles of the lower row are as stout as those of the main row, but there are three in *I. grammatica* and four in *I. obscura*".

ADDITIONAL REFERENCES

Hynes, H. B. N. & Macphee, F. M. (1970). The maxillae of the British species of *Isoperla* (Plecoptera: Perlodidae). *Proc. R. ent. Soc. Lond. (A)*, **45**, 123-124.

Zwick, P. (1967). Revision der Gattung *Chloroperla* Newmann (Plecoptera). *Mitt. schweiz. ent. Ges.* **40**, 1-26.

Zwick, P. (1973). Insecta: Plecoptera. Phylogenetisches System und Katalog. *Tierreich*, Lief. **94**, 1-465.

For a review of recent work on stoneflies, see:
Hynes, H. B. N. (1976). Biology of Plecoptera. *A. Rev. Ent.*, **21**, 135-153.

INDEX TO THE KEYS

PUBLICATIONS OF THE

FRESHWATER BIOLOGICAL ASSOCIATION

The FBA produces small (A5) booklets which provide high-quality keys
for the identification of British freshwater organisms, technical manuals
on sampling, water chemistry and statistics, selected bibliographies,
and tabulated data sets of temperature, chemical variables etc. These
Scientific and **Occasional** Publications (SPs and OPs) are A5 printed
booklets, perfect bound (SPs) or stapled (some OPs) between soft covers;
prices are generally no more than £10.00 per book, including postage
and packing. **Special** Publications (SPECs) are more expensive and have
various formats, up to A4 in size. Members of the FBA are entitled
to 25% discount on all publications. Credit card facilities
(Visa/Mastercard/Eurocard) are available.

Some examples of currently available titles are given below. A full
list and current prices may be obtained from **Dept. DWS, Freshwater
Biological Association, The Ferry House, Far Sawrey, Ambleside, Cumbria,
LA22 0LP, England.**

SCIENTIFIC PUBLICATIONS (SPs)

SP 25. Statistical Analysis

SOME METHODS FOR THE STATISTICAL ANALYSIS OF SAMPLES OF BENTHIC
INVERTEBRATES, by J.M. Elliott. Second edition, 1977. (Reprinted
with minor corrections, 1983). Pp 1-160.   ISBN 0 900386 29 0

SP 27. Freshwater Fishes

A KEY TO THE FRESHWATER FISHES OF THE BRITISH ISLES, WITH NOTES ON
THEIR DISTRIBUTION AND ECOLOGY, by P.S. Maitland, 1972. Pp 1-139.
                                             ISBN 0 900386 18 5

SP 31. Dixidae (Diptera)

A KEY TO THE LARVAE, PUPAE AND ADULTS OF THE BRITISH DIXIDAE
(DIPTERA), THE MENISCUS MIDGES, by R.H.L. Disney, 1975.
Pp 1-78.   ISBN 0 900386 23 1

SP 34. Amoebae

AN ILLUSTRATED KEY TO FRESHWATER AND SOIL AMOEBAE, WITH NOTES ON
CULTIVATION AND ECOLOGY, by F.C. Page, 1976. Pp 1-155.
                                             ISBN 0 900386 26 6

SP 35. Megaloptera and Neuroptera

A KEY TO THE LARVAE AND ADULTS OF BRITISH FRESHWATER MEGALOPTERA
AND NEUROPTERA, WITH NOTES ON THEIR LIFE CYCLES AND ECOLOGY, by
J.M. Elliott, 1977. Pp 1-52.   ISBN 0 900386 27 4

SP 36. Water Analysis

WATER ANALYSIS: SOME REVISED METHODS FOR LIMNOLOGISTS, by F.J.H.
Mackereth, J. Heron & J.F. Talling. Second impression, 1989.
Pp 1-120.   ISBN 0 900386 31 2

SP 40. Leeches

A KEY TO THE BRITISH FRESHWATER LEECHES, WITH NOTES ON THEIR LIFE
CYCLES AND ECOLOGY, by J.M. Elliott & K.H. Mann, 1979. Pp 1-72 +
1 colour plate.   ISBN 0 900386 38 X

SP 44. Diatom Frustule

A GUIDE TO THE MORPHOLOGY OF THE DIATOM FRUSTULE, WITH A KEY TO THE
BRITISH FRESHWATER GENERA, by H.G. Barber & E.Y. Haworth, 1981.
Pp 1-112.   ISBN 0 900386 42 8

SP 47. Adult Mayflies (Ephemeroptera)

A KEY TO THE ADULTS OF THE BRITISH EPHEMEROPTERA, WITH NOTES ON THEIR
ECOLOGY, by J.M. Elliott & U.H. Humpesch, 1983. Pp 1-101 + 1 plate.
                                              ISBN 0 900386 45 2

SP 48. Mosquitoes (Culicidae)

KEYS TO THE ADULTS, MALE HYPOPYGIA, FOURTH-INSTAR LARVAE AND PUPAE
OF THE BRITISH MOSQUITOES (CULICIDAE), WITH NOTES ON THEIR ECOLOGY
AND MEDICAL IMPORTANCE, by P.S. Cranston, C.D. Ramsdale, K.R. Snow
& G.B. White, 1987. Pp 1-152.   ISBN 0 900386 46 0

SP 49. Larval Mayflies (Ephemeroptera)

LARVAE OF THE BRITISH EPHEMEROPTERA: A KEY WITH ECOLOGICAL NOTES,
by J.M. Elliott, U.H. Humpesch & T.T. Macan, 1988. Pp 1-145.
                                              ISBN 0 900386 47 9

SP 50. Adult Water Bugs

ADULTS OF THE BRITISH AQUATIC HEMIPTERA HETEROPTERA: A KEY WITH
ECOLOGICAL NOTES, by A.A. Savage, 1989. Pp 1-173.
                                              ISBN 0 900386 48 7

SP 51. Case-bearing Caddis Larvae (Trichoptera)

A KEY TO THE CASE-BEARING CADDIS LARVAE OF BRITAIN AND IRELAND, by
I.D. Wallace, B. Wallace & G.N. Philipson, 1990. Pp 1-237.
                                              ISBN 0 900386 49 5

SP 52. Malacostracan Crustaceans

BRITISH FRESHWATER CRUSTACEA MALACOSTRACA: A KEY WITH ECOLOGICAL NOTES,
by T. Gledhill, D.W. Sutcliffe & W.D. Williams, 1993. Pp 1-176.
                                              ISBN 0 900386 53 3